INTRODUCTION TO
ALGEBRAIC GEOMETRY

Other books by Serge Lang, available from Addison-Wesley

Algebraic Functions, 1964

Introduction to Diophantine Approximations, 1966

Introduction to Transcendental Numbers, 1966

Rapport sur la cohomologie des groupes, 1966

The Scheer Campaign, 1967

With John Tate

Collected Papers of Emil Artin, 1965

In Preparation

Elliptic Functions, 1972

Introduction to Algebraic and Abelian Functions, 1972

INTRODUCTION TO ALGEBRAIC GEOMETRY

SERGE LANG

ADDISON-WESLEY PUBLISHING COMPANY, INC.

ADVANCED BOOK PROGRAM

READING, MASSACHUSETTS

First published, 1958, by Interscience Publishers, New York and London

Second printing, November 1964

Third printing, with corrections, July 1972, by Addison-Wesley Publishing Company, Inc., Advanced Book Program

Fourth printing, November 1973, by Addison-Wesley Publishing Company, Inc., Advanced Book Program

International Standard Book Number 0-201-04164-2
Library of Congress Catalog Number 58-8620
American Mathematical Society (MOS) Subject Classification Scheme (1970): 14 A 10

Printed in the United States of America

ISBN 0-201-04164-2
EFGHIJKLMN-CO-79876543

PREFACE

Algebraic geometry is the study of systems of algebraic equations in several variables, and of the structure which one can give to the solutions of such equations. There are four ways in which this study can be carried out: analytic, topological, algebraico-geometric, and arithmetic. It almost goes without saying that these four ways are by no means independent of each other, although each one can be pushed forward by methods appropriate to its point of view.

To use analytic and topological methods, one starts with equations whose coefficients are complex numbers. One may then consider the set of zeros of the equations as a manifold, topological, or analytic, provided one makes suitable assumptions of non-singularity.

The algebraico-geometric methods are applied in dealing with equations having coefficients in an arbitrary field, the solutions of the equations being taken to lie in its algebraic closure, or in a "universal domain." The arguments used are geometric, and are supplemented by as much algebra as the taste of the geometer will allow.

One frequently meets a problem which consists in relating invariants arising from topological, analytic, or algebraic methods. For instance, the genus of a curve may alternatively be defined as the number of holes in its Riemann surface, the number of differentials of first kind, the dimension of the Jacobian variety, or that integer which makes the Riemann-Roch formula valid. One must then prove that all these numbers are equal.

Finally, in arithmetic, one has equations whose coefficients lie in the field of rational numbers, or fields canonically derived from it. One then studies properties which depend essentially on the special nature of the coefficients field selected: algebraic number fields (finite extensions of the rationals), finite fields obtained from those by reducing modulo p, and more generally fields of finite type, generated from those by a finite number of elements, which may be transcendental. This fourth approach to algebraic equations includes of course all of diophantine analysis.

The purpose of the present book is to give a rapid, concise, and self-contained introduction to the algebraic aspects of the third approach, the algebraico-geometric, without presupposing any extensive knowledge of algebra (local algebra in particular). It is not meant as a complete treatise, but we hope that after becoming acquainted with it, the reader will find the door open to a more thorough study of the literature. With this in mind, we have appended to each chapter a short list of papers which the reader may find stimulating. We have also made some historical comments when it seemed appropriate to do so.

We have not touched any topic related to the intersection theory. This would have taken us beyond the intended scope of our book, which we hope will be used not only as an introduction to algebraic geometry, but also as an introduction to Weil's *Foundations*. There are many basic results contained in *Foundations* which still cannot be found anywhere else, especially in Chapters VII and VIII; however, we have tried to include all the results which don't pertain directly to intersection theory (i.e. all the qualitative results). A discussion of the Chow coordinates is omitted, as belonging properly to intersection theory. This theory is about to undergo an extensive recasting, because, as Weil hoped, one is now able to deal with linear equivalence and the algebraic homology ring.

We have included all the theorems of a general nature which have been used recently in laying the foundations of the theory of algebraic groups. For a bibliography, the reader is referred to the end of Chapter IX. The Riemann-Roch theorem has been given in order to provide the background for a sketch of the construction of the Jacobian variety, a typical example of complete group varieties.

The reader should keep in mind that today one can develop

a more abstract algebraic geometry than that given here. It is known as the arithmetic case, and is of great importance for number theory. For instance, an affine variety V in the arithmetic case is identified with a finitely generated ring over the ordinary integers, and the absolutely algebraic points of V are the homomorphisms of this ring into finite fields. This idea, which, as Weil has pointed out, goes as far back as Kronecker, has been rediscovered in recent times by Kähler and Weil, and given new impetus by Weil in his address to the International Congress of 1950. For a systematic attempt at laying the foundations of the arithmetic case, we refer the reader to M. Nagata, "A general theory of algebraic geometry over Dedekind domains," American Journal of Mathematics, January 1956. We have never brought out this theory explicitly here, but it is a profitable activity to translate theorems into their analogues in the arithmetic case.

S. L.

PREREQUISITES

We assume a knowledge of elementary algebra, roughly up to the level of Galois theory. We also use freely the Hilbert basis theorem (every ideal in a polynomial ring over a field has a finite set of generators), and in the chapter on normalization, we use some other simple statements on Noetherian rings. Otherwise, the book is self-contained.

The word "ring" will always mean "commutative ring with unity element and without divisors of zero" except in Chapter VIII § 1 and Chapter X.

From Chapter II on, if k is a field, then $\bar{k}$ denotes its algebraic closure.

Finally, note that we use "surjective" as the adjective form of the preposition "onto".

CONTENTS

Chapter V

Normal Varieties

Chapter VI

Divisors and Linear Systems

Chapter VII

Differential Forms

Chapter VIII

Theory of Simple Points

CHAPTER IX

Algebraic Groups

CHAPTER X

Riemann-Roch Theorem

CHAPTER I

General Theory of Places

We are concerned here with extending the ordinary theory of homomorphisms. We know that a homomorphism of a field must be trivial, i.e. must be an isomorphism. However, by generalizing the notion and allowing mappings which send some of the elements of the field onto a suitable infinity, we get "places" which prove very useful in number theory and algebraic geometry. The notion of a place was already known to Dedekind and Weber, but it is the extension theorem which gives us a powerful tool, and this has become available only comparatively recently.

If we consider an algebraic curve, then it is well known that through each point we can pass an analytic branch on the Riemann surface. The functions of the curve then have a well determined value (which may be a complex number, or ∞) at the place on the Riemann surface over the point. Furthermore the zeros and infinities of the functions can be measured by their orders.

We shall obtain a generalization of these notions, which can be formulated completely abstractly for arbitrary rings and fields. A homomorphism of a ring generalizes the notion of point on a curve, and a place of a field will generalize the place on a Riemann surface. The notion of valuation of a field will generalize the notion of order of a zero or pole.

1. Definition of places

Let F be a field. We shall work with an element ∞ satisfying the following algebraic rules. For $a \in F$ we define

$$a \pm \infty = \infty, \qquad a \cdot \infty = \infty \quad \text{if} \quad a \neq 0,$$
$$\infty \cdot \infty = \infty, \qquad 1/0 = \infty \qquad \text{and} \quad 1/\infty = 0.$$

The expressions $\infty \pm \infty$, $0 \cdot \infty$, $0/0$, and ∞/∞ are undefined.

A *place* φ of a field K into a field F is a mapping $\varphi : K \to \{F, \infty\}$

of K into the set consisting of F and ∞, satisfying the usual rules for a homomorphism namely $\varphi(a + b) = \varphi(a) + \varphi(b)$, $\varphi(ab) = \varphi(a)\varphi(b)$ whenever the expressions on the right sides of these formulas are defined, and such that $\varphi(1) = 1$. We shall also say that the place is *F-valued*. The elements of K which are not mapped into ∞ will be called *finite* under the place, and the others will be called *infinite*. If $\varphi(a) = 0$ we say that a has a zero.

Let $\mathfrak{o}$ be the set of finite elements of K under the place φ. It is clear that $\mathfrak{o}$ is a ring. If $\varphi(a) \neq 0, \infty$ then $a^{-1} \epsilon \mathfrak{o}$. Hence the image of $\mathfrak{o}$ under φ is a field. It is also clear that $\mathfrak{o}$ has a unique maximal ideal $\mathfrak{p}$ which consists of all elements of $\mathfrak{o}$ which are mapped onto 0 by the place. The field onto which $\mathfrak{o}$ is mapped is canonically isomorphic to $\mathfrak{o}/\mathfrak{p}$.

The ring $\mathfrak{o}$ satisfies the following important property: If $a \epsilon K$ and $a \notin \mathfrak{o}$, then $a^{-1} \epsilon \mathfrak{o}$. If K is a field and $\mathfrak{o}$ is a subring of that field having this property, then we shall say that $\mathfrak{o}$ is a *valuation ring of K*.

We shall now show that valuation rings give rise to places and that there is essentially a $1 - 1$ correspondence between them.

Let K be a field and $\mathfrak{o}$ a valuation ring of K. We contend that the non-units of $\mathfrak{o}$ form a maximal ideal $\mathfrak{p}$. Indeed, suppose $a, b \epsilon \mathfrak{o}$ are non-units. Say $a/b \epsilon \mathfrak{o}$. Then $1 + a/b = (a + b)/b \epsilon \mathfrak{o}$. If $a + b$ were a unit then $1/b \epsilon \mathfrak{o}$, contradicting the fact that b is not a unit. One sees trivially that for $c \epsilon \mathfrak{o}$, ca is not a unit. This proves our contention.

We consider the natural map $\mathfrak{o} \rightarrow \mathfrak{o}/\mathfrak{p}$ and put $\mathfrak{o}/\mathfrak{p} = F$. This is a field because $\mathfrak{p}$ is maximal in $\mathfrak{o}$. We define a mapping φ of K by letting φ map the elements of $\mathfrak{o}$ on their residue classes mod $\mathfrak{p}$ and by putting $\varphi(a) = \infty$ if $a \epsilon K$, but $a \notin \mathfrak{o}$. It is easily verified that φ is an *F-valued* place.

It is convenient to introduce an equivalence relation between places. Recall first that if we have two homomorphisms $\varphi_1 : R \rightarrow S_1$ and $\varphi_2 : R \rightarrow S_2$ of a ring R onto rings S_1 and S_2 respectively, then we say that these homomorphisms are *equivalent* if there exists an isomorphism $\lambda : S_1 \rightarrow S_2$ such that $\varphi_2 = \lambda\varphi_1$.

Let $\varphi_1 : K \to \{F_1, \infty\}$ and $\varphi_2 : K \to \{F_2, \infty\}$ be two places. Taking their restrictions to their images, we may assume that they are surjective. We shall say that they are *equivalent* if there exists an isomorphism $\lambda : F_1 \to F_2$ such that $\varphi_2 = \lambda\varphi_1$. (We put $\lambda(\infty) = \infty$.) One sees that two places are equivalent if and only if they have the same valuation ring.

It is clear that there is a $1 - 1$ correspondence between equivalent places and valuation rings of a field, this correspondence being given by the above constructions which are inverse to each other. A place is said to be *trivial* if it is an isomorphism. Its valuation ring is simply the whole field.

2. Valuations

Let Γ be a multiplicative commutative group. We shall say that an *ordering* is defined in Γ if we are given a subset S of Γ which is closed under multiplication (a semigroup) and such that $\Gamma = S \cup 1 \cup S^{-1}$ (disjoint), i.e. Γ is the disjoint union of S, the unit element 1, and the set S^{-1} consisting of all inverses of elements of S.

If $\alpha, \beta \in \Gamma$ we define $\alpha < \beta$ if $\alpha\beta^{-1} \in S$. We have $\alpha < 1$ if and only if $\alpha \in S$. One easily verifies the following properties of the relation $<$:

1. For any two elements α, β of Γ we have $\alpha < \beta$ or $\alpha = \beta$ or $\beta < \alpha$, and these possibilities are mutually exclusive.
2. $\alpha < \beta$ implies $\alpha\gamma < \beta\gamma$ for any $\gamma \in \Gamma$.
3. $\alpha < \beta$ and $\beta < \gamma$ implies $\alpha < \gamma$.

Conversely, a relation satisfying the three properties gives rise to a semigroup S consisting of all elements < 1. We leave the verification of this step to the reader.

It is convenient to attach to an ordered group formally an extra element 0, such that $0 \cdot \alpha = 0$, and such that $0 < \alpha$ for all $\alpha \in \Gamma$. This is useful in valuation theory. The ordered group is then analogous to the multiplicative group of the positive reals, except that it may have non-archimedean ordering.

The following proposition is trivial but useful.

PROPOSITION 1. *If $\alpha^n = 1$, then $\alpha = 1$, or $n = 0$.*

This means that the map $\alpha \to \alpha^n$ is an isomorphism of Γ into itself. Indeed, if $\alpha \neq 1$, we may assume $\alpha \in S$, and hence α^n cannot be 1 for any n, because S is a semigroup which does not contain 1.

We shall now consider valuations. A *valuation* v of a field K is a map $a \to |a| = v(a)$ of K into an ordered group Γ, with the extra element 0, such that

1. $|a| = 0$ if and only if $a = 0$.
2. $|ab| = |a| \cdot |b|$.
3. $|a + b| \leq \max(|a|, |b|)$.

Let K^* be the multiplicative group of non-zero elements of K. Then the valuation gives a homomorphism of K^* into Γ. The valuation is called *trivial* if it maps K^* into 1. If the map is not surjective, we may restrict our attention to the precise image of K, and this gives a valuation onto an ordered subgroup of Γ. If we have two valuations v_1 and v_2 of a field K, with ordered groups Γ_1 and Γ_2 both assumed to be surjective, we shall say that they are *equivalent* if there is an order-preserving isomorphism $\lambda : \Gamma_1 \to \Gamma_2$ such that $v_2 = \lambda v_1$.

Valuations have the following additional properties:

4. $|1| = 1$, because $|1| = |1|^2$.
5. $|\pm a| = |a|$.   (The proof is obvious.)
6. If $|a| < |b|$ then $|a + b| = |b|$.

Proof: Under our hypothesis, we have
$$|b| = |b + a - a| \leq \max(|b + a|, |a|) = |b + a| \leq$$
$$\max(|b|, |a|) = |b|.$$

7. In a sum $a_1 + a_2 + \ldots + a_n = 0$, at least two elements of the sum have the same value.

This is an immediate consequence of the preceding property.

We shall now connect the valuations with valuation rings and places.

Let φ be a place of a field K, and $\mathfrak{o}$ its valuation ring, $\mathfrak{p}$ the maximal ideal of $\mathfrak{o}$. Let U be the units of $\mathfrak{o}$. We can write $K^* = \mathfrak{p}^* \cup U \cup \mathfrak{p}^{*-1}$. Here, $\mathfrak{p}^*$ means the non-zero elements of $\mathfrak{p}$ and the union is disjoint. Consider the factor group K^*/U and denote a coset aU by $|a|$. We also put $|0| = 0$. We define an ordering in the group K^*/U by putting $|a| < 1$ if and only if $a \in \mathfrak{p}^*$. The set S of values $|a| < 1$ is clearly a semigroup and if we let $\Gamma = K^*/U$ then $\Gamma = S \cup 1 \cup S^{-1}$ (disjoint).

Furthermore we note that $|a| < |b| \Longleftrightarrow |a/b| < 1 \Longleftrightarrow a/b \in \mathfrak{p}$ $\Longleftrightarrow \varphi(a/b) = 0$. We obviously have

$$|ab| = |a||b| \quad \text{and} \quad |a + b| \leq \max\,(|a|, |b|).$$

Hence the place gives rise to a valuation of K if we put $|0| = 0 < |a|$ for all $a \neq 0$.

Conversely, let Γ be an ordered group and $|\ |$ a valuation of K in Γ. Let $\mathfrak{o}$ be the subset of K consisting of all elements a such that $|a| \leq 1$. Then $\mathfrak{o}$ is a valuation ring. If $|a| = 1$ then $|a^{-1}| = 1$ also. In that case a is a unit of $\mathfrak{o}$, and is also called a *unit* for the valuation. From this we see that the maximal ideal $\mathfrak{p}$ of $\mathfrak{o}$ consists of all $a \in \mathfrak{o}$ such that $|a| < 1$.

We see that there is a $1 - 1$ correspondence between equivalent valuations, equivalent places, and valuation rings.

3. Examples

We shall determine all places of the rationals.

Let $\mathbf{Q}$ be the rational numbers and $\mathbf{Z}$ the integers. Let φ be a non-trivial place of $\mathbf{Q}$. Then one sees immediately that φ cannot be trivial on the integers. Since $\varphi(1) = 1$, it follows that φ is finite on $\mathbf{Z}$. The intersection of the prime ideal $\mathfrak{p}$ with Z is a prime ideal of $\mathbf{Z}$ which is not 0, because we assumed φ is non-trivial. All prime ideals of $\mathbf{Z}$ are of type (p) where p is a prime number. We have $\varphi(p) = 0$. If m is an integer, $p \nmid m$, then $\varphi(m) \neq 0$ and is finite. In fact φ induces a homomorphism on Z, namely $\mathbf{Z} \to \mathbf{Z}/p\mathbf{Z}$, and $\varphi(m)$ is the residue class mod p. If we write a rational number in the form $p^r m/n$, where $p \nmid mn$, then

$$\varphi(p^r m/n) = \begin{cases} \varphi(m)/\varphi(n) & \text{if } r = 0 \\ 0 & \text{if } r > 0 \\ \infty & \text{if } r < 0. \end{cases}$$

The value group associated with this place is infinite cyclic. The valuation ring consists of all fractions whose donominator is not divisible by p, and its maximal ideal consists of those fractions whose numerator is divisible by p.

We shall determine all places φ of a rational field $k(x) = K$ (where x is transcendental over k) which are trivial on k.

Let I be the ring of polynomials $k[x]$. Suppose φ is non-trivial. Then φ is non-trivial on I. If $\varphi(x) = \infty$ then we replace x by $1/x$, and consider the ring $k[1/x]$. Hence we may assume without loss of generality that φ is finite on x and hence finite on I. The intersection of the zeros $\mathfrak{p}$ of φ with I is a prime ideal which is principal, of type $p(x)$, where $p(x)$ is a polynomial irreducible over k. Hence φ maps I on a ring isomophic to the residue class ring I/pI, in analogy with the integers. If we write a rational function as $p^r m/n$, where m, n are polynomials prime to $p = p(x)$, then we have the same sort of situation that we had for the rationals. In this case φ maps rational functions on their function values. The residue class ring I/pI is isomorphic to a finite algebraic extension of k, namel $k(\theta)$, where θ is a root of $p(x)$. The function values lie in $k(\theta)$. If k is algebraically closed, then of course an irreducible polynomial is linear, and may be chosen of type $(x - \theta)$ with $\theta \in k$, and φ is k-valued, sending x onto θ.

More generally, consider the case where R is an arbitrary unique factorization domain. Let p be a prime of R. An element f of the quotient field can be written $f = p^r g/h$, where g and h are elements of R not divisible by p. Then a place, valuation ring, and valuation can be defined on the quotient field K in exactly the same manner as was done before in the case of the rational numbers or of a function field in one variable. An important example of a unique factorization domain is given by the polynomial ring in several variables $R = k[x_1, \ldots, x_n]$, where the x_i are algebraically independent over a field k. The units of R are then the elements of k, and the primes are simply the irreducible polynomials $p = p(x_1, \ldots, x_n)$ over k.

All the value groups arising from the preceding examples are infinite cyclic. They are very important, but there are other value groups. We leave it to the reader to prove that given an ordered group there exists a field having a valuation whose value group is the given one.

The method of constructing such a field is illustrated by the following special case. Let A be the ordered group (written additively) generated over the integers by 1 and $\sqrt{2}$. Take a rational field $k(x, y)$ in two variables. Substitute formally $y = x^{\sqrt{2}}$ in a rational function $f(x, y)/g(x, y)$. Then setting $x = 0$, we see that we can compute the value of a rational function in an obvious manner. Our place gives us a transcendental parametrization in a neighborhood of the origin $(0, 0)$ in the plane.

4. Extension of places

The theorem proved in this section is fundamental. It is used constantly, and makes possible the simplification of many proofs in algebraic geometry.

Let us start with a definition. By a *local ring* $\mathfrak{o}$ one means a ring with a unique maximal ideal $\mathfrak{m}$. The elements of $\mathfrak{o}$ which are not in $\mathfrak{m}$ must then be the units of $\mathfrak{o}$.

Now let R be a ring and $\mathfrak{p}$ a prime ideal of R. We consider the set $R_{\mathfrak{p}}$ consisting of all elements a/z with $a, z \in R$ but $z \notin \mathfrak{p}$. It is immediately verified to be a local ring, which is called the *local ring of R at $\mathfrak{p}$*. Its maximal ideal consists of all quotients a/z with a in $\mathfrak{p}$ and $z \notin \mathfrak{p}$. A homomorphism of a local ring $\mathfrak{o}$ whose kernel is the maximal ideal $\mathfrak{m}$ of $\mathfrak{o}$ is said to be *at the center of $\mathfrak{o}$*. The canonical homomorphism $\mathfrak{o} \to \mathfrak{o}/\mathfrak{m}$ is at the center of $\mathfrak{o}$.

If φ is a homomorphism of R into a field, say an algebraically closed field Ω, then the kernel of φ is a prime ideal $\mathfrak{p}$, and the local ring $R_{\mathfrak{p}}$ is said to be the *local ring of φ*.

If K is a field containing R, we are concerned with the possibility of extending φ to an Ω-valued place of K. We shall begin by showing how φ extends in a natural way to a homomorphism of the local ring $R_{\mathfrak{p}}$. First note that if φ is so extended to a homomorphism φ^*, then $\varphi^*(a/z) = \varphi(a)/\varphi(z)$, so the extension is unique.

To define the extension, given an element a/z of $R_{\mathfrak{p}}$, we let $\varphi(a/z) = \varphi(a)/\varphi(z)$. This is well defined because $a/z = a'/z' \Rightarrow az' = za' \Rightarrow \varphi(a)\varphi(z') = \varphi(z)\varphi(a') \Rightarrow \varphi(a)/\varphi(z) = \varphi(a')/\varphi(z')$. This extension of φ to $R_{\mathfrak{p}}$ is then trivially verified to be a homomorphism. Thus we have shown that a homomorphism of a ring R into Ω can always be extended to a homomorphism at the center of the local ring $R_{\mathfrak{p}}$, and in a unique manner.

If φ is trivial on R, then its kernel is zero, and the local ring is the quotient field K of R. We see that the extension of φ is trivial on K. More generally, we have:

PROPOSITION 2. *Let R be a ring, φ_0 an isomorphism of R, L a field containing R and algebraic over the quotient field K of R. Then every place φ of L extending φ_0 is trivial on L.*

Proof: If φ is not trivial on L, then there exists an element y of L which is mapped onto 0 by φ, and $y \neq 0$. We have an equation

$$y^n + a_{n-1}y^{n-1} + \ldots + a_0 = 0$$

with $a_i \epsilon K$, and $a_0 \neq 0$. Applying φ, we get $0 = \varphi(a_0)$ which contradicts the hypothesis that φ is trivial on K.

We now come to our fundamental theorem.

THEOREM 1. *Let K be a field containing a ring R, Let φ be a homomorphism of R into an algebraically closed field Ω, and suppose $\varphi(1) = 1$. Then φ can be extended to an Ω-valued place of K.*

Proof: In view of the discussion at the beginning of this section, we may assume that R is a local ring $\mathfrak{o}$ and that φ is at the center of $\mathfrak{o}$ i.e. that the kernel of φ is the maximal ideal $\mathfrak{m}$ of $\mathfrak{o}$. The image of $\mathfrak{o}$ under φ is a subfield F of Ω.

It will be shown that given an element x of K, the homomorphism φ can be extended to a homomorphism of the ring $\mathfrak{o}[x]$ or $\mathfrak{o}[x^{-1}]$ into Ω.

LEMMA. *Let $\mathfrak{o}$ be a local ring contained in a field K, and let $\mathfrak{m}$ be its maximal ideal. Let x be an element of K, and consider the ideal generated by $\mathfrak{m}$ in the rings $\mathfrak{o}[x]$ and $\mathfrak{o}[x^{-1}]$ respectively. Then it cannot happen that in both cases the ideal is equal to the whole ring.*

Proof: Suppose it did happen. We could then write an equation

(1) $1 = p_0 + p_1 x + \ldots + p_n x^n,$

with p_i in the maximal ideal $\mathfrak{m}$ of $\mathfrak{o}$. We assume that n is the smallest degree of such equations for x. Similarly, we would have

$$(2) \qquad 1 = q_0 + q_1 x^{-1} + \ldots + q_m x^{-m}$$

with $q_j \in \mathfrak{m}$, and we assume that the degree m is minimal for such equations for x^{-1}.

Say $m \leq n$. We multiply (2) by x^m. We then obtain an expressio for x^m as a polynomial in x of degree $< m$, and with coefficients in $\mathfrak{m}$. Substituting this expression in (1), we obtain an equation for x of the same type, but of lower degree. This is a contradiction and proves the lemma.

Say that the ideal $\mathfrak{m}\mathfrak{o}[x]$ is not the whole ring. Then it is contained in a maximal ideal $\mathfrak{P}$ of $\mathfrak{o}[x]$. Furthermore $\mathfrak{P} \cap \mathfrak{o}$ contains $\mathfrak{m}$ and hence must be $\mathfrak{m}$ since $\mathfrak{m}$ is maximal in $\mathfrak{o}$. Let σ be a homomorphism of $\mathfrak{o}[x]$ with kernel $\mathfrak{P}$ into some field. Then the restriction of σ to $\mathfrak{o}$ is equivalent to φ, i.e. they have the same kernel $\mathfrak{m}$. Without loss of generality we may assume that the restriction of σ to $\mathfrak{o}$ is equal to φ, so $\sigma\mathfrak{o} = \varphi\mathfrak{o} = F$. If $\sigma(x) = t$, then t may be transcendental over F, and $\sigma(\mathfrak{o}[x]) = F[t]$. We now take a homomorphism τ of $F[t]$ into Ω mapping t on an element of Ω algebraic over F, and such that τ is identity on F. This can be done because Ω is assumed algebraically closed. Then the composite homomorphism $\tau \circ \sigma$ gives us the desired extension of φ to $\mathfrak{o}[x]$.

We can apply Zorn's lemma to finish the proof of the extension theorem. We consider the set of all pairs (S, ψ) consisting of a subring S of K containing R, and a homomorphism ψ of S into Ω whose restriction to R is φ. These pairs can be inductively ordered by prescribing $(S, \psi) \leq (S', \psi')$ if $S \subset S'$ and the restriction of ψ' to S is ψ. If (S, ψ) is a maximal element then S must be a local ring and ψ must be at the center of S, and, moreover, the preceding arguments show that S must in fact be a valuation ring. We can then extend ψ to a place by giving it the value infinity on all elements of K which are not in S. This concludes the proof of the theorem.

The reader will note that we have not merely proved that a homomorphism can be extended to a place, but that it can be extended to a place taking its values in the given algebraically closed field Ω.

For some applications it is necessary to complete the extension theorem by giving more precise information of a quantitative nature concerning the extension of the place. This information is contained in Theorem 2. If φ is a homomorphism of a ring R, and $f(X) = a_0 + a_1 X + \ldots + a_n X^n$ is a polynomial with coefficients in R, then we denote by $\bar{f}$ the polynomial obtained by applying φ to all the coefficients of f, so

$$\bar{f}(X) = \varphi(a_0) + \varphi(a_1)X + \ldots + \varphi(a_n)X^n = \bar{a}_0 + \bar{a}_1 X + \ldots + \bar{a}_n X^n.$$

THEOREM 2. *Let R be a ring, K its quotient field, and φ a homomorphism of R into an algebraically closed field Ω. Let $f(X) = a_n X^n + \ldots + a_0$ be a polynomial with coefficients in R, of degree n, and let $(x_1, \ldots, x_n)$ be its roots in an algebraic closure of K, each counted with its multiplicity. Assume that $\bar{f}$ is not identically 0, and let m be the degree of $\bar{f}$. Then for any extension φ^* of φ to a place of $K(x_1, \ldots, x_n)$, the set $(\varphi^*(x_1), \ldots, \varphi^*(x_n))$ is uniquely determined up to a permutation, and consists of all the roots of $\bar{f}$, each taken with its multiplicity, and of the element ∞ which appears $n - m$ times.*

Proof: Given a place φ^* of $K(x_1, \ldots, x_n)$ extending φ, we denote by $\bar{y}$ the image under φ^* of an element y of $K(x_1, \ldots, x_n)$. After a suitable renumbering of the x_i, we may assume that $\bar{x}_1, \ldots, \bar{x}_r \neq \infty$, and $\bar{x}_{r+1}, \ldots, \bar{x}_n = \infty$. We can write

$$f(X) = y(X - x_1) \ldots (X - x_r)\left(\frac{1}{x_{r+1}}X - 1\right) \ldots \left(\frac{1}{x_n}X - 1\right),$$

where $y = a_n x_{r+1} \ldots x_n$. Then $\varphi^*(y)$ can be neither 0 nor ∞. Indeed, if $\varphi^*(y) = \infty$, we divide both sides by y and apply φ^*. Then the left-hand side would give 0, while the right-hand side would not, and this would be a contradiction. If $\varphi^*(y) = 0$, then we would have $\bar{f} = 0$, contrary to assumption.

The above arguments show that $\bar{f}$ is equal to

$$\bar{y}(X - \bar{x}_1) \ldots (X - \bar{x}_r)(- 1)^{n-r}$$

and we see therefore that $\bar{x}_1, \ldots, \bar{x}_r$ are precisely the roots of f. This concludes the proof of our theorem.

Frequently the polynomial $f(X)$ is irreducible over K, and its roots $(x_1, \ldots, x_n)$ are then called a complete set of conjugates over K. Each one appears with a multiplicity which is equal to a power of the characteristic. From the fact that they are all conjugate over K, we have the following corollary.

COROLLARY. *Assume in the preceding theorem that $f(X)$ is irreducible over K. Let x be one of its roots. Then there exists an extension φ^* of φ to $K(x)$ such that $\varphi^*(x) = \infty$ if and only if deg $\bar{f} < \deg f$.*

Proof: We know from the theorem that if the degree of $\bar{f}$ is smaller than that of f, then one of the roots of f goes to infinity under a place extending φ. There is an automorphism of the algebraic closure of K mapping x on this root. If we follow this automorphism by the place, we get what we want.

Suppose that x is an element of some field containing R. If $f(X)$ is a polynomial in $R[X]$ such that $f(x) = 0$ but $\bar{f}$ is not identically zero, then any Ω-valued place φ^* of $K(x)$ extending φ must map x on ∞ or on some root of $\bar{f}$. In particular, there is only a finite number of possible values $\varphi^*(x)$. If, on the other hand, for every polynomial $f(X)$ in $R[X]$ such that $f(x) = 0$ we have $\bar{f} = 0$, then we can pick an arbitrary element t of Ω and φ can be extended to a homomorphism φ^* of $R[x]$ by putting $\varphi^*(x) = t$, and more generally $\varphi^*(g(x)) = \bar{g}(t)$ for $g \in R[X]$. In particular, t can be selected transcendental over R. Summarizing, we get

PROPOSITION 3. *Let R be a ring, φ a homomorphism into an algebraically closed field F, and x an element of a field L containing R. If there are infinitely many extensions of φ to F-valued places φ_i of L such that the $\varphi_i(x)$ are all distinct, then there exists an extension of φ to a place φ^* of L such that $\varphi^*(x)$ is finite and transcendental over F. If in particular x is transcendental over the quotient field of R, then such a place exists.*

5. Integral closure

Let K be a field and R a subring. Classically, an element x of K is said to be *integral over* R if it satisfies an equation

$$x^n + a_{n-1}x^{n-1} + \ldots + a_0 = 0$$

with a_i in R and leading coefficient 1. Such an equation will be called an *integral equation for x over R*. We shall give a characterization of integral elements in terms of places.

PROPOSITION 4. *Let K be a field containing a ring R. An element x of K is integral over R if and only if every place of K finite on R is finite on x.*

Proof: Suppose x satisfies an equation of the above type, and φ is a place of K finite on R. If $\varphi(x) = \infty$ divide the equation by x^n and apply φ. This gives $1 = 0$, a contradiction.

Assume that every place φ finite on R is finite on x. Consider the ring $S = R[1/x]$. (We may clearly assume $x \neq 0$.) If $1/x$ is a unit in this ring then we can write

$$x = a_0 + a_1(1/x) + \ldots + a_{n-1}(1/x^{n-1})$$

with $a_i \in R$. Multiply by x^{n-1}. This gives an integral equation for x over R.

If $1/x$ is not a unit in S, then $1/x$ generates a proper principal ideal. By Zorn's lemma, this ideal is contained in a maximal ideal $\mathfrak{p}$. The homomorphism $\varphi : S \to S/\mathfrak{p}$ can be extended to a place of K, and this place is finite on R but maps $1/x$ on 0. Hence the place is infinity on x, a contradiction. This proves our proposition.

Our characterization of integral elements in terms of places immediately shows that the set of elements x of K which are integral over R form a ring, which is called the *integral closure* of R in K. If K is the quotient field of R and R is equal to its integral closure in K, then we shall say that R is *integrally closed*. The integral closure of R in K is obviously the intersection of all valuation rings of K containing R.

As consequences of our characterization of integral elements, we obtain some corollaries.

COROLLARY 1. *If we have three rings $R_1 \subset R_2 \subset R_3$ such that R_2 is integral over R_1 and R_3 is integral over R_2 then R_3 is integral over R_1.*

This principle of transitivity follows immediately from the proposition.

COROLLARY 2. *Let $S = R[x_1, \ldots, x_n]$ be a ring finitely generated over a ring R. If each x_i is integral over R, and if φ is a homomorphism of R into some algebraically closed field Ω, then there is only a finite number of extensions of φ to a homomorphism of S into Ω.*

Proof: An integral equation for x_i over R will be mapped under φ into a non-trivial equation which must be satisfied by an image of x_i under an extension of φ to S, and such an equation has only a finite number of roots in Ω.

Let R be a ring contained in another ring S. Let $\mathfrak{a}$ be an ideal of R. An ideal $\mathfrak{A}$ of S will be said to *lie above* $\mathfrak{a}$ if $\mathfrak{A} \cap R = \mathfrak{a}$.

COROLLARY 3. *If $\mathfrak{p}$ is a prime ideal of a ring R, and S is a ring containing R and integral over R, then there exists always a prime ideal $\mathfrak{P}$ of S lying above $\mathfrak{p}$.*

Proof: This is equivalent to saying that a homomorphism of R into an algebraically closed field can always be extended to a homomorphism of S into that field.

COROLLARY 4. *The assumptions being as in Corollary 3, the residue class ring $S/\mathfrak{P}$ is algebraic over $R/\mathfrak{p}$. If $\mathfrak{p}$ is maximal, so is $\mathfrak{P}$.*

Proof: Here we have identified $R/\mathfrak{p}$ in $S/\mathfrak{P}$, and one sees that every element of $S/\mathfrak{P}$ is algebraic over $R/\mathfrak{p}$ by looking at an integral equation for an element of S over R. If $\mathfrak{v}$ is maximal, then $R/\mathfrak{p}$ is a field, and $S/\mathfrak{P}$ must be a field also, hence $\mathfrak{P}$ is maximal.

COROLLARY 5. *Let R be a ring contained in a ring S, integral over R. Let $\mathfrak{p}$ be a prime ideal of R, and $\mathfrak{P}$, $\mathfrak{Q}$ two prime ideals of S lying above $\mathfrak{p}$. If $\mathfrak{P} \subset \mathfrak{Q}$, then $\mathfrak{P} = \mathfrak{Q}$.*

Proof: There is a canonical homomorphism of $S/\mathfrak{P}$ onto $S/\mathfrak{Q}$ which is identity on $R/\mathfrak{p}$. Since both $S/\mathfrak{P}$ and $S/\mathfrak{Q}$ are algebraic over $R/\mathfrak{p}$, this homomorphism must be an isomorphism (Prop. 2) and hence $\mathfrak{P} = \mathfrak{Q}$.

One can often test whether an element is integral over a ring R by using only a subset of all places finite on R. We shall give two examples of this situation.

PROPOSITION 5. *Let R be a ring, $\mathfrak{p}$ a prime ideal, and φ a homomorphism with kernel $\mathfrak{p}$. Let K be a field containing R. An element x of K is integral over the local ring $R_\mathfrak{p}$ if and only if x is finite under every place of K extending φ, or equivalently at every place of K at the center of $R_\mathfrak{p}$.*

Proof: The last equivalence is due to the uniqueness of the extension of φ to the local ring $R_\mathfrak{p}$.

If x is integral over $R_\mathfrak{p}$ then we know that every place finite on $R_\mathfrak{p}$ is finite on x. Conversely, let $\mathfrak{m}$ be the maximal ideal of $R_\mathfrak{p}$, and put $R_\mathfrak{p} = \mathfrak{o}$. If $1/x$ is a unit in $\mathfrak{o}[1/x]$, then it is integral over $\mathfrak{o}$ by an argument similar to the one used in Proposition 4. If the ideal $(\mathfrak{m}, 1/x)$ of $\mathfrak{o}[1/x]$ is the entire ring, then we can write

$$-1 = a_n(1/x)^n + \ldots + a_1(1/x) + y$$

where y is an element of $\mathfrak{m}$ and $a_i \in \mathfrak{o}$. From this we get

$$x^n(1 + y) + \ldots + a_n = 0.$$

But $1 + y$ is not in $\mathfrak{m}$ and hence is a unit in $\mathfrak{o}$. We can divide the equation by $1 + y$ to conclude that x is integral over $\mathfrak{o}$. If finally the ideal $(\mathfrak{m}, 1/x)$ is not the entire ring, then it is contained in a maximal ideal, whose intersection with $\mathfrak{o}$ contains $\mathfrak{m}$ and hence must be equal to $\mathfrak{m}$. We would then obtain a homomorphism at the center of $\mathfrak{o}$ which maps $1/x$ on 0. Extending this to a place would give a contradiction.

The second example which we had in mind is that of a unique factorization domain.

PROPOSITION 6. *Let R be a unique factorization domain. Let K be its quotient field, and let x be algebraic over K. Let*

$$a_n x^n + \ldots + a_0 = 0$$

be an irreducible equation for x over K, with coefficients a_i in R, *and assume that the a_i are relatively prime, i.e. no prime p of R divides all a_i. Then x is integral over R if and only if a_n is a unit in R.*

Proof: If a_n is a unit we can divide the equation by a_n to get an integral equation for x over R. If a_n is not a unit we can find a prime p of R dividing a_n. Let σ be the canonical homomorphism $R \to R/(p)$. By assumption $\sigma(a_i) \neq 0$ for all i. Using the corollary of Theorem 2, we see that σ can be extended to a place of $K(x)$ which is infinity on x, whence x cannot be integral over R.

6. Places in algebraic extensions

The study of the behaviour of places in an algebraic extension of a given field K is a very natural one. We give here only a few basic and elementary results concerned with it. More such results will be found in the chapter on the normalization of varieties.

Let L be an arbitrary field containing a field K. Let v be a valuation of L, with value group Γ_L and units U_L. Then v induces a valuation on K, and we may view the value group of this induced valuation as a subgroup of Γ_L. On the other hand, we may also view it as $\Gamma_K = K^*/U_K$. There is a natural isomorphism of K^*/U_K into Γ_L which is order-preserving. Indeed, we have $U_K = U_L \cap K^*$. Hence the map

$$K^*/U_K \to L^*/U_L$$

given by $aU_K \to aU_L$ is an isomorphism, which is obviously order-preserving. We therefore identify Γ_K in Γ_L.

PROPOSITION 7. *Let L/K be a finite algebraic extension of degree n. Then the index $(\Gamma_L : \Gamma_K)$ is finite and $\leq n$.*

Proof: Let $y_1, \ldots, y_r$ be elements in L representing distinct cosets of Γ_K in Γ_L. We prove that the y_i are linearly independent over K. In a relation $a_1 y_1 + \ldots + a_r y_r = 0$ with $a_i \in K$, $a_i \neq 0$, two terms must have the same absolute value, say $|a_i y_i| = |a_j y_j|$ (Property 7 of valuations), and hence $|y_i| = |a_i^{-1} a_j||y_j|$. This contradicts the assumption that y_i, $y_j (i \neq j)$ represent distinct cosets and proves the proposition.

COROLLARY 1. *If Γ_K is infinite cyclic, then so is Γ_L.*

Proof: If e is the index in our proposition, then every element of Γ_L raised to the eth power lies in Γ_K. Hence raising to the eth power gives an isomorphism of Γ_L onto a subgroup of Γ_K (Proposition 1). This proves the corollary.

A valuation whose value group is infinite cyclic is said to be *discrete*. We conclude from the corollary that all the valuations of an algebraic number field (a finite extension of the rationals) are discrete, because we have seen previously in § 3 that all the valuations of the rationals are discrete. In the chapter on divisors, we shall again meet discrete valuations. We note here already that when we deal with a discrete valuation of a field K, we are able to speak of the order of the zero or pole of an element of K. Indeed, the value group is canonically isomorphic to the integers, a canonical generator being given by the value $|\pi|$ of an element π of K such that π lies in the valuation ring, and $|\pi|$ generates the value group. Given an element x of K, we can find a power of π such that $x = \pi^r u$, where u is a unit in the valuation ring. If for instance $r > 0$, we say that x has a zero of order r.

We have another finiteness statement dealing with the residue class fields. Let L be a field containing another field K, and let $\mathfrak{O}$ be a valuation ring of L, with maximal ideal $\mathfrak{P}$. Let $\mathfrak{o} = \mathfrak{O} \cap K$. Then $\mathfrak{o}$ is a valuation ring of K. Let $\mathfrak{p}$ be its maximal ideal. Then the residue class ring $\mathfrak{o}/\mathfrak{p}$ has a canonical isomorphism in $\mathfrak{O}/\mathfrak{P}$, and we usually identify $\mathfrak{o}/\mathfrak{p}$ in $\mathfrak{O}/\mathfrak{P}$.

PROPOSITION 8. *Let L be a finite extension of degree n of a field K, and let $\mathfrak{O}$ be a valuation ring of L. Let $\mathfrak{P}$ be its maximal ideal, let $\mathfrak{o} = \mathfrak{O} \cap K$, and $\mathfrak{p}$ be the maximal ideal of $\mathfrak{o}$. Then the residue class degree $[\mathfrak{O}/\mathfrak{P} : \mathfrak{o}/\mathfrak{p}]$ is finite and $\leq n$.*

Proof: Select $y_1, \ldots, y_r$ in $\mathfrak{O}$ such that their residue classes mod $\mathfrak{P}$ are linearly independent over $\mathfrak{o}/\mathfrak{p}$. Then in a relation $\sum a_i y_i = 0$, $a_i \in K$, $a_i \neq 0$, we may assume all a_i to be integers (i.e. elements of $\mathfrak{o}$) and some a_i to be a unit (if necessary divide the equation by the coefficient having the biggest value in the associated valuation). Reducing the equation mod $\mathfrak{P}$ shows that the y_i must have been linearly independent over $\mathfrak{o}/\mathfrak{p}$, and proves our proposition.

We are interested in proving that given a valuation ring $\mathfrak{o}$ in a field K, and a finite algebraic extension L of K, then there is only a finite number of valuation rings $\mathfrak{O}$ in L such that $\mathfrak{O} \cap K = \mathfrak{o}$. Such valuation rings will be said to *lie above* $\mathfrak{o}$ in L. Actually, we can prove a more general statement concerning integrally closed rings. First we need a statement concerning integral elements.

PROPOSITION 9. *Let x be integral over a ring R and let K be the quotient field of R. Then all conjugates of x over K are integral over R. If R is integrally closed then the irreducible equation of x over K with leading coefficient 1 has all its coefficients in R.*

Proof: If x' is a conjugate of x over K, then there exists an isomorphism of $K(x)$ over K mapping x on x'. Applying this isomorphism to an integral equation for x over R shows that x' is also integral over R. Since the coefficients of the irreducible equation for x are polynomial functions of the roots, they are integral over R. They are also in K, and if R is integrally closed, they are in R. This proves our proposition.

THEOREM 3. *Let R be a ring, K its quotient field, and $\mathfrak{p}$ a prime ideal of R. Assume that R is integrally closed. Let L be a normal extension of K, and S the integral closure of R in L. Let $\mathfrak{P}$ and $\mathfrak{O}$ be two prime ideals of S lying above $\mathfrak{p}$. Then there exists an automorphism σ of L/K such that $\mathfrak{O} = \mathfrak{P}^\sigma$.*

Proof: Let $\mathfrak{P}_i (i = 1, \ldots, m)$ be all the distinct prime ideals of S which are conjugate to $\mathfrak{P}$, i.e. of type $\mathfrak{P}^\sigma$ for some σ in the group of automorphisms G of L/K. Suppose $\mathfrak{O}$ is not among the $\mathfrak{P}_i$. Then the product $\mathfrak{O}\mathfrak{P}_1 \ldots \hat{\mathfrak{P}}_i \ldots \mathfrak{P}_m$ with $\mathfrak{P}_i$ omitted cannot be contained in $\mathfrak{P}_i$, according to Corollary 5 of Proposition 4 of § 5. Hence we can find an element x_i in this product, but $x_i \notin \mathfrak{P}_i$. Let $y = x_1 + \ldots + x_m$. Then one sees immediately that y is in $\mathfrak{O}$ but y is not in any $\mathfrak{P}_i$. If a conjugate y^σ of y lies in some $\mathfrak{P}_j$, then y lies in $\mathfrak{P}_j^{\sigma^{-1}}$ which is among the $\mathfrak{P}_i$. Hence y^σ cannot be in any $\mathfrak{P}_j$. The norm of y, which is equal to $(\prod_\sigma y^\sigma)^{p^s}$ (p being the characteristic) cannot be in any $\mathfrak{P}_j$ either. However it is in S (by Proposition 9) and in K, hence in R because R is integrally closed. It is also in $\mathfrak{O}$ (because y itself enters in the product and $\mathfrak{O}$

is an ideal). Being in $\mathfrak{Q} \cap R$, it is in $\mathfrak{p}$ which is contained in each $\mathfrak{P}_i$. This is a contradiction, and proves the theorem.

COROLLARY. *Let L/K be a finite algebraic extension, and let R be a subring of K, integrally closed and having K as quotient field. Let $\mathfrak{p}$ be a prime ideal of R. Let S be a subring of L containing R and integral over R. Then there is only a finite number of prime ideals in S which lie above $\mathfrak{p}$.*

Proof: If E is the smallest normal extension of K containing L and S^* the integral closure of S (or of R) in E, then the preceding theorem gives us the finiteness statement in E. It is now clear that it must also hold in the subfield L of E.

THEOREM 4. *Let R be a valuation ring of a field K. Let L/K be an algebraic extension. Let $\mathfrak{O}$ be a valuation ring of L lying above R, and $\mathfrak{M}$ its maximal ideal. Let S be the integral closure of R in L, and let $\mathfrak{P} = \mathfrak{M} \cap S$. Then $\mathfrak{O}$ is equal to the local ring $S_{\mathfrak{P}}$.*

Proof: It is clear that $S_{\mathfrak{P}}$ is contained in $\mathfrak{O}$. Conversely, let x be an element of $\mathfrak{O}$. Then x satisfies an equation with coefficients in K:

$$a_n x^n + \ldots + a_0 = 0, \qquad a_i \, \epsilon \, K.$$

Suppose that a_s has the biggest value among the a_i for the valuation associated with the valuation ring, and that it is the coefficient furthest to the left having this value. Let $b_i = a_i/a_s$. Then all b_i are in R, and $b_n, \ldots, b_{s+1}$ are in $\mathfrak{M}$. Divide the equation by x. We get

$$(b_n x^{n-s} + \ldots + b_{s+1} x + 1) + \frac{1}{x}\left(b_{s-1} + \ldots + b_0 \frac{1}{x^{s-1}}\right) = 0.$$

Let y and z be the two quantities in parentheses in the preceding equation, so we can write

$$y = \frac{1}{x}z \quad \text{and} \quad xy = z.$$

To prove our theorem it will suffice to show that y and z are in S and that y is not in $\mathfrak{P}$.

Taking into account Proposition 5 of § 5, we let φ be a place

CHAPTER II

Algebraic Varieties

We introduce the geometric language, and define the notion of algebraic sets and varieties. It is convenient to develop the theory relative to a fixed ground field k, and later to study the technique of constant field extensions.

The main theorems of this chapter are the Hilbert Nullstellensatz and the dimension theorem. The former characterizes an algebraic set by its algebraic zeros, and the latter asserts that the dimension of every component of the intersection of two varieties V, W of dimensions r, s respectively in affine space S^n has dimension at least $r + s - n$ provided of course that the intersection is not empty. This dimension theorem is used very frequently in later chapters.

We have also included the classical elimination theory, which will be seen to be a direct consequence of the extension theorem proved in Chapter I. Here we study essentially projective varieties, principally to guarantee that certain intersections of varieties are not empty.

1. Notation and preliminaries

We choose in this section some notation which will be used throughout the rest of this book.

We let Ω be a field of infinite degree of transcendence over the prime field, and algebraically closed. We call Ω the *universal domain*. The field of complex numbers is such a field, for instance. If K is a subfield of Ω, then $\bar{K}$ will always denote its algebraic closure in Ω.

We consider a family of subfields K, L, . . ., of Ω such that
1. Ω has infinite degree of transcendence over them;
2. the prime field is in our family and if K is in the family so is

every field obtained from K by adjoining a finite number of elements of Ω to K;

3. if K and L are in the family so is the compositum KL (the compositum being the smallest subfield of Ω containing both K and L).

Unless otherwise specified, a field will always mean a member of that family.

By n-space S^n we understand the Cartesian product of Ω n times: $S^n = \Omega \times \ldots \times \Omega$. A *point* in n-space, or, briefly, a point, is an n-tuple $(x_1, \ldots, x_n)$ with components $x_i \in \Omega$. We denote elements of Ω by small letters, x, y, $\ldots$, often with subscripts. We call them *quantities*. We use vector notation, and write (x) instead of $(x_1, \ldots, x_n)$.

If k is a field we denote the ring of formal polynomials in n variables over k by $k[X] = k[X_1, \ldots, X_n]$. Its quotient field is $k(X)$.

If (x) is a set of quantities (or a point) we denote the ring of polynomials $k[x_1, \ldots, x_n]$ by $k[x]$ and its quotient field by $k(x)$.

Throughout the present chapter, we shall work over a fixed field k in our family, and the theory to be developed can be said to be relative to k. It will be convenient to use an absolute theory later, and this will be discussed in the next chapter.

For want of a better place, we insert here an elementary algebraic result, known as the Noether normalization theorem.

THEOREM 1. *Let $k[x]$ be a finitely generated ring over a field k, and assume that $k(x)$ has transcendence degree r over k. Then there exist elements $y_1, \ldots, y_r$ in $k[x]$ such that $k[x]$ is integral over $k[y]$.*

Proof: If the x's are already algebraically independent over k, we are done. If not, there is a non-trivial algebraic relation

$$\sum c_{(j)} x_1^{j_1} \cdots x_n^{j_n} = 0,$$

where each coefficient $c_{(j)}$ is in k, and $c_{(j)} \neq 0$. The sum is taken over a finite number of distinct vectors $(j_1, \ldots, j_n)$ of integers $j_\nu \geqq 0$. Let $m_2, \ldots, m_n$ be positive integers, and put

$$y_2 = x_2 - x_1^{m_2}, \ldots, y_n = x_n - x_1^{m_n}.$$

Substitute $x_i = y_i + x_1^{m_i}$ $(i = 2, \ldots, n)$ in the above equation.

Using vector notation, we put $(m) = (1, m_2, \ldots, m_n)$ and use the dot product $(j) \cdot (m)$ to denote $j_1 + m_2 j_2 + \ldots + m_n j_n$. If we expand the relation after making the above substitutions, we get

$$\sum c_{(j)} x_1^{(j) \cdot (m)} + f(x_1, y_2, \ldots, y_n) = 0,$$

where f is a polynomial in which no pure power of x_1 appears.

We shall prove that for suitable (m) all the integers $(j) \cdot (m)$ are distinct. Once this is done, we will have a relation of integral dependence for x_1 over $k[y_2, \ldots, y_n]$. It is then clear by transitivity that $k[x]$ is integral over $k[y_2, \ldots, y_n]$ and we can then proceed inductively.

There are many ways of showing that $(j) \cdot (m)$ are distinct for suitable (m). One of them runs as follows. Since we have assumed all the vectors (j) distinct, their mutual differences are non-zero. Let $d^{(1)}, \ldots, d^{(s)}$ be these differences. We can write $d^{(\nu)} = (d_{\nu 1}, \ldots, d_{\nu n})$. Let $T_2, \ldots, T_n$ be variables and consider the polynomial

$$F(T_2, \ldots, T_n) = \prod_{\nu=1}^{s} (d_{\nu 1} + T_2 d_{\nu 2} + \ldots + T_n d_{\nu n}).$$

Then $F(T)$ is not identically zero, and hence there exist (infinitely many) values $(m_2, \ldots, m_n)$ such that $F(m_2, \ldots, m_n) \neq 0$; this is easily proved by induction on the number of variables. Such values $(m_2, \ldots, m_n)$ fulfill our requirements.

If the field k is infinite then there is a more linear way of constructing the elements $y_1, \ldots, y_r$, and it is sometimes useful to have the following result.

THEOREM 1'. *Let k be a field, and $k[x] = k[x_1, \ldots, x_n]$. Suppose that $k(x)$ over k has transcendence degree r. Let $(u) = (u_{ij})$ $(i = 1, \ldots, r$ and $j = 1, \ldots, n)$ be rn quantities algebraically independent over $k(x)$, and put*

$$y_i = \sum_{\nu=1}^{n} u_{i\nu} x_\nu \qquad\qquad i = 1, \ldots, r.$$

Let $K = k(u)$. Then $K[x]$ is integral over $K[y]$.

The proof, which is included in most books, will here be left to the reader.

2. Decomposition theorem

Let $\mathfrak{a}_0$ be an ideal in the polynomial ring $k[X]$. A point (x) is called a *zero* of $\mathfrak{a}_0$ if $f(x) = 0$ for all $f \epsilon \mathfrak{a}_0$. The set A of all zeros of $\mathfrak{a}_0$ is called an *algebraic set* determined by $\mathfrak{a}_0$. (Later, it will also be called *a k-closed set*. See Chapter III.)

Let $\mathfrak{a}$ be the set of all polynomials $g \epsilon k[X]$ vanishing on A, i.e. such that $g(x) = 0$ for all $(x) \epsilon A$. Then $\mathfrak{a}$ is an ideal in $k[X]$ which is said to *belong to* A. Clearly $\mathfrak{a} \supset \mathfrak{a}_0$, but the converse is not always true. The conditions under which $\mathfrak{a} = \mathfrak{a}_0$ will be discussed later (Hilbert Nullstellensatz).

We see immediately that A is also determined by the ideal $\mathfrak{a}$. Indeed, let A' be determined by $\mathfrak{a}$. Then $\mathfrak{a} \supset \mathfrak{a}_0$ implies $A' \subset A$, and the converse inclusion $A \subset A'$ is trivial. We say that $\mathfrak{a}$ and A *belong to each other*, or that $\mathfrak{a}$ is *associated with* A.

If A and B are algebraic sets and $\mathfrak{a}$, $\mathfrak{b}$ their associated ideals then it is clear that $A \subset B$ if and only if $\mathfrak{a} \supset \mathfrak{b}$. Hence $A = B$ if and only if $\mathfrak{a} = \mathfrak{b}$. This has a very important consequence, namely, that the ideals of $k[X]$ satisfy the ascending chain condition. We can therefore conclude that algebraic sets satisfy the descending chain condition: A chain $A_1 \underset{\neq}{\supset} A_2 \underset{\neq}{\supset} A_3 \underset{\neq}{\supset} \ldots$ must stop.

THEOREM 2. *The finite union* $\overset{m}{\underset{i=1}{\cup}} A_i$ *and the finite intersection* $\overset{m}{\underset{i=1}{\cap}} A_i$ *of algebraic sets are algebraic sets. In fact, if A and B are two algebraic sets and $\mathfrak{a}$, $\mathfrak{b}$ their associated ideals, then $\mathfrak{a} \cap \mathfrak{b}$ belongs to $A \cup B$ and $(\mathfrak{a}, \mathfrak{b})$ belongs to $A \cap B$.*

Proof: We may clearly restrict ourseives to two algebraic sets A and B. We prove first that $A \cup B$ is an algebraic set, and is in fact determined by the product $\mathfrak{a}\mathfrak{b}$. Let (x) be in $A \cup B$. Then trivially (x) is a zero of $\mathfrak{a}\mathfrak{b}$. Conversely, let (x) be a zero of $\mathfrak{a}\mathfrak{b}$, and (x) not in A. There exists a polynomial $f \epsilon \mathfrak{a}$ such that $f(x) \neq 0$. But $f\mathfrak{b} \subset \mathfrak{a}\mathfrak{b}$, and hence $(fg)(x) = 0$ for all g in $\mathfrak{b}$, whence $g(x) = 0$ for all $g \epsilon \mathfrak{b}$. This proves that (x) is in $\mathfrak{b}$, and that $A \cup B$ is an algebraic set. One sees immediately that a polynomial vanishes on $A \cup B$ if and only if it is in $\mathfrak{a} \cap \mathfrak{b}$, so $\mathfrak{a} \cap \mathfrak{b}$ is the associated ideal of $A \cup B$.

To prove $A \cap B$ is determined by $(\mathfrak{a}, \mathfrak{b})$, let $(x) \in A \cap B$. Then (x) is a zero of $(\mathfrak{a}, \mathfrak{b})$. Conversely, let (x) be a zero of $(\mathfrak{a}, \mathfrak{b})$. Then obviously $(x) \in A \cap B$, as desired. It is now obvious that $(\mathfrak{a}, \mathfrak{b})$ is actually the ideal associated with $A \cap B$.

An algebraic set V is called *k-variety* if it cannot be expressed as a proper union of two algebraic sets, i.e. if $V \neq A \cup B$, with A and B distinct from V. In this chapter we call V a variety.

THEOREM 3. *Let A be an algebraic set. Then A can be expressed as a finite union of varieties*: $A = V_1 \cup \ldots \cup V_r$. *If there is no inclusion relation among the V_i, i.e. if $V_i \not\subset V_j$ $(i \neq j)$ then this representation is unique.*

Proof: We first show existence, which is a consequence of the chain condition. Let $A = B \cup C$, $B \neq A$ and $C \neq A$. Let $\mathfrak{a}, \mathfrak{b}, \mathfrak{c}$ belong to A, B, C respectively. Then $\mathfrak{a} \gneq \mathfrak{b}$ and $\mathfrak{a} \gneq \mathfrak{c}$. If A could not be expressed as a finite union of varieties, then we could go down an infinite descending tree of algebraic sets, contradicting the chain condition.

Let us prove the uniqueness under the assumption that there is no inclusion relation among the V_i. Let $A = \cup V_i = \cup W_j$, where the V_i, W_j are varieties. For each W_j we can write

$$W_j = (W_j \cap V_1) \cup \ldots \cup (W_j \cap V_r).$$

Since each $W_j \cap V_i$ is an algebraic set, we must have $W_j = V_i \cap W_j$ for some i. Hence $W_j = V_i$ for some i. Similarly, V_i is contained in W_ν. Since there is no inclusion relation among the W's we must have $W_j = V_i = W_\nu$. This argument can be carried out for each W and each V. This proves that each W appears among the V's and each V appears among the W's, and proves our representation is unique.

THEOREM 4. *An algebraic set A is a variety if and only if its associated ideal is prime.*

Proof: Let V be a variety and let $\mathfrak{p}$ be its associated ideal. If $\mathfrak{p}$ is not prime, we can find two polynomials $f, g \in k[X]$ such that $f \notin \mathfrak{p}, g \notin \mathfrak{p}$, but $fg \in \mathfrak{p}$. Let $\mathfrak{a} = (\mathfrak{p}, f)$ and $\mathfrak{b} = (\mathfrak{p}, g)$. Let A be determined by $\mathfrak{a}$ and B be determined by $\mathfrak{b}$. Then $A \subset V$, $A \neq V$

and $B \subset V$, $B \neq V$. Furthermore, $A \cup B = V$. Indeed, $A \cup B \subset V$ trivially. Conversely, let $(x) \in V$. Then $(fg)(x) = 0$ implies $f(x)$ or $g(x) = 0$. Hence $(x) \in A$ or $(x) \in B$, proving $V = A \cup B$.

Let V be an algebraic set and let its associated ideal $\mathfrak{p}$ be prime. Suppose $V = A \cup B$, $A \neq V$ and $B \neq V$. Let $\mathfrak{a}$, $\mathfrak{b}$ belong to A and B. There exists a polynomial $f \in \mathfrak{a}$, $f \notin \mathfrak{p}$ and a polynomial $g \in \mathfrak{b}$, $g \notin \mathfrak{p}$. But fg vanishes on $A \cup B$ and hence lies in $\mathfrak{p}$, contradiction.

Whenever an algebraic set is expressed as a union of varieties V_i without inclusion relations, then we shall call these varieties its *k-components*. (In this chapter, we shall say *components*). Similarly we shall call *k*-varieties simply varieties here. In all chapters except this one we shall never omit the *k*.

3. Generic points and specializations

Let $\mathfrak{a}$ be an ideal of $k[X]$ and not equal to $k[X]$. In order to find a zero of $\mathfrak{a}$ it suffices to find a zero of a prime ideal containing $\mathfrak{a}$. Such ideals exist, say by Zorn's lemma. We shall therefore investigate prime ideals.

Let $\mathfrak{p}$ be a prime ideal of $k[X]$. The canonical homomorphism

$$k[X] \to k[X]/\mathfrak{p}$$

induces an isomorphism on k. We shall identify k with its image. If we denote by ξ_i the image of X_i under the map, it follows that $k[X]/\mathfrak{p}$ is isomorphic to the ring $k[\xi_1, \ldots, \xi_n]$. We have $f \in \mathfrak{p}$ if and only if $f(\xi) = 0$, because $\mathfrak{p}$ is the exact kernel of the map $f(X) \to f(\xi)$.

LEMMA. *Let $k(\xi)$ be a finitely generated extension of k. There exists an isomorphism of $k(\xi)$ into Ω which is identity on k.*

Proof: Say $\xi_1, \ldots, \xi_r$ is a transcendence basis for $k(\xi)$ over k. Let $x_1, \ldots, x_r$ be algebraically independent elements of Ω over k. There is an isomorphism of $k(\xi_1, \ldots, \xi_r)$ onto $k(x_1, \ldots, x_r)$ mapping each ξ_i on x_i, and this isomorphism can be extended to the algebraic extension $k(\xi)$ of $k(\xi_1, \ldots, \xi_r)$.

All the natural maps to be considered, will be taken to be identity on k, unless otherwise specified.

By the lemma, we can map the ring $k[X]/\mathfrak{p}$ isomorphically into Ω, $k[\xi] \cong k[x]$ where (x) is a set of quantities. For each prime ideal $\mathfrak{p}$ of $k[X]$ we can therefore find a point (x) such that $k[X]/\mathfrak{p}$ is canonically isomorphic to $k[x]$ under the map $f(X) \to f(x)$ and $f(x) = 0$ if and only if $f \in \mathfrak{p}$. We call (x) a *generic zero of* $\mathfrak{p}$.

If (x') is another point, and if there is an isomorphism $\varphi : k[x] \to k[x']$ sending $f(x)$ onto $f(x')$ then it is clear that $f(x) = 0$ if and only if $f(x') = 0$. Hence both rings $k[x]$ and $k[x']$ are isomorphic to $k[X]/\mathfrak{p}$.

The previous discussion shows that a prime ideal determines a class of points (x), $\mathfrak{p}$ being the kernel of the homomorphism $f(X) \to f(x)$.

Conversely, given a point (x), the set of all polynomials $f \in k[X]$ such that $f(x) = 0$ is obviously an ideal $\mathfrak{p}$. It is prime because $(fg)(x) = 0$ implies $f(x) = 0$ or $g(x) = 0$, and hence $f \in \mathfrak{p}$ or $g \in \mathfrak{p}$. This prime ideal is the kernel of the map $f(X) \to f(x)$. We say that $\mathfrak{p}$ is associated with (x), or that $\mathfrak{p}$ is the *ideal determined by* (x) (over k). We see that (x) is a generic zero of $\mathfrak{p}$.

Let (x') be a point. Let $\mathfrak{p}'$ be determined by (x'). We say that (x') is a *specialization of* (x) *over* k if $\mathfrak{p} \subset \mathfrak{p}'$, i.e. if every polynomial vanishing on (x) also vanishes on (x'). We write $(x) \to (x')$, omitting the reference to k whenever it is clear over which field we are taking the specialization.

Let $(x) \to (x')$ be a specialization. Then the map $k[x] \to k[x']$ sending $f(x)$ onto $f(x')$ is well defined because by hypothesis, $f(x) = 0$ implies $f(x') = 0$. It is then obviously a homomorphism. Conversely, if (x) and (x') are two points, and if the map $f(x) \to f(x')$ is a homomorphism of $k[x]$ onto $k[x']$ then $(x) \to (x')$ is a specialization.

If $k[x]$ and $k[x']$ are isomorphic under the map $f(x) \to f(x')$ we say that (x') is a *generic specialization of* (x), or that (x) and (x') are equivalent (over k, of course).

Let V be a variety, and $\mathfrak{p}$ its associated prime ideal. Let (x) be a generic zero of $\mathfrak{p}$. A point (x') is in V if and only if it is a specialization of (x). Indeed, let $\mathfrak{p}'$ be the prime ideal determined by (x'). If (x') is in V, then $\mathfrak{p}$ vanishes on (x') and consequently

$\mathfrak{p} \subset \mathfrak{p}'$, and $(x) \to (x')$ is a specialization. Conversely, if $(x) \to (x')$ is a specialization, we have $\mathfrak{p} \subset \mathfrak{p}'$ by definition and consequently (x') is in V.

We shall call (x) a *generic point of V*. We see immediately that if (x) is a generic point of V and (x') is a generic specialization of (x), then (x') is also a generic point of V. Every point (x) may therefore be looked at as a generic point of some variety. Indeed, the set of all specializations of (x) is a variety V, which consists precisely of the zeros of the prime ideal $\mathfrak{p}$ determined by (x). We shall call this variety the *locus of (x) over k*, or simply the locus of (x) when the reference to k is clear. We also say that V is a *model* of the field $k(x)$. Any finitely generated field over k has a model, and in fact many models. A set of generators can always be looked upon as the generic point of such a model.

The ring $k[x]$ is called a *coordinate ring* of the variety V, and from the above discussion one sees that the points of V are in $1 - 1$ correspondence with the homomorphisms of the coordinate ring into the universal domain.

We have associated a prime ideal with a variety. Conversely, given a prime ideal $\mathfrak{p}$, let V be the algebraic set of zeros of $\mathfrak{p}$. Let (x) be a generic zero of $\mathfrak{p}$. If f is a polynomial vanishing on V, then in particular $f(x) = 0$. Hence $f \in \mathfrak{p}$. This proves that $\mathfrak{p}$ is the ideal associated with V, and consequently that V is a variety (by Theorem 4).

The preceding discussion has established a natural correspondence between the following objects:

1. Points (x) which are generic specializations of each other;
2. prime ideals $\mathfrak{p}$;
3. varieties.

We define the *dimension* of a point (x) over k to be the transscendence degree of $k(x)$ over k. We write $\dim_k(x)$, or, briefly, $\dim(x)$ when the reference to k is clear.

If V is a variety and (x) a generic point of V, we let $\dim V = \dim(x)$. The dimension of V thus defined is clearly independent of the generic point selected. To express the fact that a variety has dimension r, we sometimes write V^r.

THEOREM 5. *Let (x') be a specialization of (x) over k. Then* $\dim (x') \leqq \dim (x)$, *and equality holds if and only if (x) and (x') are generic specializations of each other.*

Proof: If $x'_1, \ldots, x'_r$ are algebraically independent over k, then the inverse images $x_1, \ldots, x_r$ of the map $f(x) \to f(x')$ must also be independent because by definition of a specialization, $f(x) = 0$ implies $f(x') = 0$. This proves the inequality.

Suppose that the dimensions of (x) and (x') are equal. Since (x') is a specialization of (x) we have a natural homomorphism $\varphi : k[x] \to k[x']$ and we must show that φ is an isomorphism. If it is not, then there exists an element y in the ring $k[x]$ such that $\varphi(y) = 0$. Say $x'_1, \ldots, x'_r$ is a transcendence base of $k(x')$. Then $x_1, \ldots, x_r$ must be a transcendence base of $k(x)$ by assumption on the dimension and the preceding paragraph. Hence y satisfies an equation

$$a_n y^n + \ldots + a_0 = 0$$

with $a_j \epsilon k[x_1, \ldots, x_r]$ and $a_0 \neq 0$. Applying φ to this equation gives a contradiction because φ is an isomorphism on $k[x_1, \ldots, x_r]$ and therefore $\varphi(a_0) \neq 0$.

COROLLARY 1. *Let $V \supset W$ be two varieties. Then* $\dim W \leq \dim V$ *and equality holds if and only if $V = W$.*

Proof: Let $\mathfrak{p}$ and $\mathfrak{p}'$ be the prime ideals associated with V and W in $k[X]$. Then $\mathfrak{p} \subset \mathfrak{p}'$. The first statement is clear if we look at the generic points (x) and (x') of V and W and apply the theorem. If $\dim (x) = \dim (x')$ then $k[x]$ is isomorphic to $k[x']$, whence $\mathfrak{p} = \mathfrak{p}'$. This shows that $V = W$. The converse is obvious.

COROLLARY 2. *Two points are equivalent if and only if they are specializations of each other.*

Proof: If they are specializations of each other their dimensions must be equal, and hence the specializations are generic, by the theorem. The converse is obvious.

Let A be an algebraic set. We know that $A = \cup V_i$ is a finite union of varieties, and that these varieties are uniquely determined. This means that the points of A consist of the specializations of a finite number of generic points (x), each (x) belonging to one of the varieties. This remark gives the following corollary.

COROLLARY 3. *A point (x') of A is the generic point of one of the components of A if and only if there is no point (x) of A, of higher dimension than (x') such that $(x) \to (x')$ over k.*

Proof: Let (x') be a generic point of a component V of A. If $(x) \to (x')$ over k, and if dim $(x) >$ dim (x') for some $(x) \in A$, then the variety W belonging to (x) contains V properly, by Corollary 1. Since the varieties in a decomposition of A have no inclusion relation this is impossible.

Conversely let (x') lie in A. Then (x') is in some V. It is a specialization of the generic point of V. If we assume that (x') has maximal dimension, we apply Theorem 5 to prove that (x') is generic for V.

In the preceding discussions, we have dealt with varieties in S^n and their associated prime ideals in $k[X]$. The following generalization is sometimes useful.

Given a variety V, and (x) a generic point, we consider an ideal $\mathfrak{a}_0$ in the coordinate ring $k[x]$. A point (x') is called a zero of $\mathfrak{a}_0$ if for each $f(x)$ in $\mathfrak{a}_0$ we have $f(x') = 0$. In other words, $f(x)$ becomes zero under the homomorphism $\varphi : k[x] \to k[x']$. Just as for ideals in $k[X]$, we see immediately that the set of zeros of $\mathfrak{a}_0$ is an algebraic set A contained in V. In fact, if we let $\mathfrak{a}_0^*$ be the ideal in $k[X]$ consisting of all polynomials $f(X)$ such that $f(x)$ is in $\mathfrak{a}_0$, then we see first that $\mathfrak{a}_0^*$ contains the prime ideal associated with V in $k[x]$, and that A is the algebraic set of zeros of $\mathfrak{a}_0^*$. Furthermore, the set of all elements $f(x)$ in $k[x]$ such that f vanishes on A is again an ideal $\mathfrak{a}$ in $k[x]$ which is said to be the ideal *associated with A in $k[x]$*.

If A happens to be a subvariety of V, then its associated ideal in $k[x]$ is obviously a prime ideal, and one sees immediately that if (x') is a generic point of A, then this prime ideal consists of all elements $f(x)$ in $k[x]$ such that $f(x') = 0$. It is said to be the *ideal determined by (x') in $k[x]$*. Conversely, one also sees that every prime ideal $\mathfrak{p}$ of $k[x]$ is associated with a subvariety of V, and that it is the set of all elements $f(x)$ of $k[x]$ vanishing on V.

It is trivially verified that if A and B are two algebraic sets contained in the variety V, and if $\mathfrak{a}$ and $\mathfrak{b}$ are their associated ideals in the coordinate ring $k[x]$ of V, then $A \subset B$ if and only if $\mathfrak{a} \supset \mathfrak{b}$ and hence $A = B$ and if only if $\mathfrak{a} = \mathfrak{b}$.

We end this section with some remarks on the local ring of a point. Let R be the coordinate ring $k[x]$ of a variety V, and let $\mathfrak{p}$ be a prime ideal of R. Let W be the subvariety of V determined by $\mathfrak{p}$, and (x') a generic point of W. The local ring $R_{\mathfrak{p}}$ defined in Ch. I. § 4 consists of all quotients $f(x)/g(x)$ with $f(x), g(x)$ in R and $g(x) \notin \mathfrak{p}$, or equivalently $g(x') \neq 0$. This local ring will be called the *local ring of* (x') *in* $k(x)$ or *of* W *in* $k(x)$.

An element z of the field $k(x)$ is said to be *defined* at a point (x') if there exists a representation of z as the quotient $f(x)/g(x)$ of two elements of $k[x]$ such that $g(x') \neq 0$, or in other words if z is in the local ring of (x') in $k(x)$. One could also say that z is *holomorphic* at (x'). We say that z is *defined at a subvariety* W of V if z is defined at a generic point of W, and we say that z is *everywhere defined* on V if it is defined at all points of V.

(By the way, the use of the word holomorphic in the above connection cannot lead to confusion. It would be possible to define holomorphic by considering the completion of the local ring, but one can prove that an element z of $k(x)$ holomorphic in such an analytic sense must necessarily be in the local ring.)

THEOREM 6. *Let* (x) *be a generic point of a variety* V. *If an element* z *of* $k(x)$ *is everywhere defined on* V *then* z *is in the coordinate ring* $k[x]$. (*The converse is obvious.*)

Proof: For each point (x') of V we can write $z = f(x)/g(x)$ with $g(x') \neq 0$. The set of all denominators $g(x)$ arising in this fashion generates an ideal in $k[x]$. This ideal cannot be proper, for otherwise it would be contained in a maximal ideal, which would induce a common zero of all the denominators, contradicting the hypothesis that z is everywhere defined. Hence the unit element 1 lies in the ideal, and we can write

$$1 = h_1(x)g_1(x) + \ldots + h_m(x)g_m(x).$$

We have $g_i(x)z = f_i(x)$. Hence

$$\sum h_i(x)g_i(x)z = z$$

lies in $k[x]$, as was to be shown.

4. The Hilbert Nullstellensatz

The theorems of this section are immediate consequences of the extension theorem of Chapter I.

THEOREM 7. *Let (x) be a point and k a field. Then (x) has an algebraic specialization over k. In other words, there exists a point (x') such that $k(x')$ is algebraic over k and such that $(x) \to (x')$ is a specialization.*

Proof: Say $x_1, \ldots, x_r$ is a transcendence base of $k(x)$ over k. Each x_i satisfies an algebraic equation of type

$$g_m(x_1, \ldots, x_r)x_i^m + \ldots + g_0(x_1, \ldots, x_r) = 0.$$

Let $(x_1', \ldots, x_r')$ be quantities in the algebraic closure $\bar{k}$ of k such that the leading coefficients $g_m(x_1', \ldots, x_r')$ do not vanish. Let $\varphi : k[x_1, \ldots, x_r] \to k[x_1', \ldots, x_r']$ be the natural homomorphism Extend φ to a place of $k(x)$ into $\bar{k}$. Then φ cannot map any of the x_i on infinity. (Otherwise, divide the equation by x_i^m and apply φ to give a contradiction.) Let $x_i' = \varphi(x_i)$ for $i = 1, \ldots, n$. Then (x') is an algebraic specialization of (x), as desired.

COROLLARY. *If an ideal $\mathfrak{a}$ of $k[X]$ is not the unit ideal, i.e. if $\mathfrak{a} \neq k[X]$ then $\mathfrak{a}$ has an algebraic zero.*

Proof: Imbed $\mathfrak{a}$ in a maximal ideal $\mathfrak{p}$, which exists by Zorn's lemma. Let (x) be a generic zero of $\mathfrak{p}$. Any specialization of (x) will be a zero of $\mathfrak{p}$, and *a fortiori* a zero of $\mathfrak{a}$. By the theorem there exists an algebraic specialization of (x), and this proves the corollary.

REMARK. We leave it as an exercise to the reader to verify that if $\mathfrak{p}$ is maximal, the generic zero is already algebraic over k.

The above corollary is known as the weak form of the Hilbert Nullstellensatz. A well-known trick of Rabinowitsch allows one to deduce its strong form given below as Theorem 8, but another trick will serve the same purpose.

Lemma. *Let R be a ring, and C a subset of R not containing 0, and closed under multiplication. If $\mathfrak{p}$ is an ideal of R, and is maximal subject to the condition that its intersection with C is empty, then $\mathfrak{p}$ is a prime ideal.*

Proof: Suppose a, $b \in R$, neither a nor b are in $\mathfrak{p}$, but $ab \in \mathfrak{p}$. Consider the ideal $(\mathfrak{p}, a)$ generated by $\mathfrak{p}$ and a in R. From the maximality of $\mathfrak{p}$, this ideal must meet C and we can write $\pi + wa = c$ for some π in $\mathfrak{p}$, w in R, and c in C. Similarly, we have $\pi' + w'b = c'$. Multiplying the right sides and the left sides of those two equations we get $cc' \in \mathfrak{p}$, a contradiction.

Theorem 8. *Let (x) be the generic point of a variety V and $k[x] = R$ the coordinate ring. Let $\mathfrak{a}$ be an ideal of $k[x]$, and f an element of R. If f vanishes on the algebraic zeros of $\mathfrak{a}$ then some power f^r of f lies in $\mathfrak{a}$. In particular, if $\mathfrak{a}$ is a prime ideal $\mathfrak{p}$, then $\mathfrak{p}$ is determined by its algebraic zeros.*

Proof: Consider the set C of all powers of f. It is closed under multiplication. We may assume f is not 0. If no power of f is in $\mathfrak{a}$ then by Zorn's lemma there exists a prime ideal $\mathfrak{p}$ containing $\mathfrak{a}$ and not meeting C. If (x') is a generic zero of $\mathfrak{p}$ (i.e. we have $k[x]/\mathfrak{p} \cong k[x']$) then $f(x')$ cannot be zero. Thus we have found one zero of $\mathfrak{p}$, and hence of $\mathfrak{a}$, which is not zero of f.

To find an algebraic zero, we consider the point $(x', 1/f(x'))$, and take some algebraic specialization $(x'', 1/f(x''))$ of it, using Theorem 7. Then (x'') is an algebraic zero of $\mathfrak{a}$, and $f(x'') \neq 0$. This proves the theorem.

Of course, we may take V to be S^n, in which case a coordinate ring $k[x]$ of V is isomorphic to the polynomial ring $k[X]$; actually it is this case which is Hilbert's original theorem. As a corollary, we shall also mention one more result.

Corollary. *Let (x) be the generic point of a variety V, and $k[x]$ the coordinate ring. If $\mathfrak{a}$ and $\mathfrak{b}$ are two ideals of $k[x]$ and if every element of $\mathfrak{a}$ vanishes on the algebraic set of zeros of $\mathfrak{b}$, then some power $\mathfrak{a}^r$ of $\mathfrak{a}$ is contained in $\mathfrak{b}$.*

Proof: The ideal $\mathfrak{a}$ has a finite basis, and we can apply the theorem to each element of that basis. Taking a sufficiently high exponent r, we see that $\mathfrak{a}^r$ must be contained in $\mathfrak{b}$.

5. Decompostion over the algebraic closure of k

Let k be our ground field and $\bar{k}$ be its algebraic closure. Let V be the variety of zeros of a prime ideal $\mathfrak{p}$ in $k[X]$ and let (x) be a generic point of V over k, so that $k[X]/\mathfrak{p} \cong k[x]$. The ideal $\mathfrak{p}\bar{k}[X]$ generated by $\mathfrak{p}$ in the ring $\bar{k}[X]$ is not necessarily a prime ideal. Its algebraic set of zeros is V again, but if we consider V as an algebraic set over $\bar{k}$ then we have a decomposition of V into $\bar{k}$-varieties V_j according to Theorem 3 of § 2: $V = \cup V_j$.

THEOREM 9. *Let V_j be one of the components of V over $\bar{k}$ and let $(\bar{x})$ be a generic point of V_j over $\bar{k}$. Then (x) and $(\bar{x})$ are equivalent over k, i.e. there is an isomorphism $k[x] \to k[\bar{x}]$ and so $(\bar{x})$ is a generic point of V over k. Hence $\dim_{\bar{k}} V_j = \dim_k V$.*

Proof: Since $(\bar{x})$ is in V_j, it must be in V. Hence $(x) \to (\bar{x})$ is a specialization over k. Let $\varphi : k[x] \to k[\bar{x}]$ be the corresponding homomorphism. We shall eventually prove that it is an isomorphism, i.e. that the specialization is generic. Extend φ to a place of $\bar{k}(x)$. This yields a homomorphism

$$\bar{k}[x] \to \bar{k}[\bar{x}]$$

whose restriction to $\bar{k}$ is an isomorphism, but not necessarily identity. Denote this restriction by σ. Extend σ to an isomorphism of $\bar{k}[x]$ onto some ring $\bar{k}[y]$. Then $\sigma^{-1} : \bar{k}[y] \to \bar{k}[x]$ is an isomorphism, and the composition of the maps $\varphi\sigma^{-1}$ as indicated:

$$\bar{k}[y] \overset{\sigma^{-1}}{\to} \bar{k}[x] \overset{\varphi}{\to} \bar{k}[\bar{x}]$$

is identity on $\bar{k}$. This implies that $(\bar{x})$ is a specialization of (y) over $\bar{k}$. Since (x) and (y) are equivalent over $\bar{k}$ by construction they have the same dimension, over k or $\bar{k}$. By Theorem 5, it follows that if $(\bar{x})$ is a generic point of V_j, it must have the same dimension as (y), and hence is equivalent to (y) over $\bar{k}$, as was to be shown.

It follows from our discussion that given two components V_1 and V_2 with generic points $(x)_1$ and $(x)_2$ over k, these must be conjugate, i.e. there exists an isomorphism σ such that $\sigma V_1 = V_2$ and $\sigma : k(x)_1 \to k(x)_2$.

The whole subject of isomorphisms of varieties will be taken up in considerable detail in Chapter III, where we deal with k-closed sets and the rationality of a cycle.

6. Product varieties over an algebraically closed field

As in the preceding section, we prove a result which will be generalized later, in Chapter IV. We shall use this next theorem, however, in the dimension theorem of the following section, and since the proof is very simple, we give it here.

We suppose that k is algebraically closed. Let $\mathfrak{p}$ be the ideal of a variety V in $k[X]$ and $\mathfrak{q}$ the ideal of a variety W in $k[Y]$, so V is in S^n and W in S^m. Then it is clear that $V \times W$ is the algebraic set of zeros of $(\mathfrak{p}, \mathfrak{q})$ in the ring $k[X, Y]$. Let $\mathfrak{P}$ be the ideal in $k[X, Y]$ associated with $V \times W$. It will be proved in Chapter IV that $\mathfrak{P}$ is actually equal to $(\mathfrak{p}, \mathfrak{q})$, but here we shall prove only that $V \times W$ is actually a variety, i.e. that $\mathfrak{P}$ is prime.

We know from the Hilbert Nullstellensatz that V and W are determined by their algebraic points, and we shall assume tacitly that the points used in the proof below have their coordinates in k.

THEOREM 10. *If V and W are varieties and k is algebraically closed, and $\mathfrak{P}$ is the ideal associated with $V \times W$ in $k[X, Y]$ as above, then $\mathfrak{P}$ is a prime ideal, and $V \times W$ is therefore a variety.*

Proof: Let $f(X, Y)g(X, Y)$ be in $\mathfrak{P}$, i.e. vanish on $V \times W$. We must show that f or g lies in $\mathfrak{P}$.

Let $\mathfrak{a}$ be the ideal generated by all polynomials $f(X, y)$ for all (y) in W. Then $\mathfrak{a}$ is in $k[X]$.

Let $\mathfrak{b}$ be the ideal generated by all polynomials $g(X, y)$ for all (y) in W. Then $\mathfrak{b}$ is in $k[X]$.

We contend that $\mathfrak{ab} \subset \mathfrak{p}$ (the ideal of V in $k[X]$). If this is not the case, there exists (x) in V and (y), (y') in W such that $f(x, y)g(x, y') \neq 0$.

We must have $f(x, y) \neq 0$, and hence $f(x, Y) \notin \mathfrak{q}$ (the ideal of W in $k[Y]$). But by assumption, $f(x, Y)g(x, Y)$ vanishes on W.

Hence $g(x, Y)$ must be in q. This implies that $g(x, y') = 0$, a contradiction.

We have proved that $\mathfrak{ab} \subset \mathfrak{p}$. Since $\mathfrak{p}$ is prime, say $\mathfrak{a} \subset \mathfrak{p}$. Then for all (x) in V and all (y) in W we have $f(x, y) = 0$. This proves that $f \in \mathfrak{P}$ vanishes on $V \times W$ and our theorem is proved.

REMARK. If we take the product of two k-varieties V and W when k is not algebraically closed, then $V \times W$ is not necessarily a k-variety. If however $V = \cup V_\alpha$ and $W = \cup W_\beta$ are the decompositions of V and W over $\bar{k}$ into $\bar{k}$-varieties, then $V \times W = \cup V_\alpha \times W_\beta$. Hence by the result of the preceding section, we get:

COROLLARY. *If V and W are two k-varieties, $r = \dim V$ and $s = \dim W$, then the dimension of each k-component of the algebraic set $V \times W$ is $r + s$.*

7. The dimension theorem

In affine n-space S^n, the algebraic set of zeros of an ideal generated by a single polynomial $f(X)$ is called a *hypersurface* H_f. It consists of all points (x) such that $f(x) = 0$. Let V be a variety of dimension r, and assume that H_f does not contain V, or equivalently that f does not vanish on all of V. Then all components of $H_f \cap V$ have dimension $\leq r - 1$ (by Corollary 1 of Theorem 5, § 3). The main result of this chapter is that all components have exactly that dimension, provided the intersection is not empty.

As a matter of notation, let $k[x]$ be the coordinate ring of V, and let z be an element of $k[x]$. Then z can be written $z = f(x)$ where $f(X)$ is in $k[X]$ and conversely every polynomial $f(X)$ in $k[X]$ determines an element $z = f(x)$ of $k[x]$. We shall sometimes call the intersection of V with this H_f the *intersection of V with the hypersurface $z = 0$*, or the *hypersurface section $z = 0$* of V.

THEOREM 11. *Let V be an affine variety of dimension r in S^n. Let $f(X)$ be a polynomial in $k[X]$. If $V \cap H_f$ is not empty, and f does not vanish on all of V, then all components of this intersection have dimension $r - 1$.*

Proof: We begin by proving the dimension theorem for the special case that $V = S^n$ is the entire space. We split $f(X)$ into irreducible factors in $k[X]$:

$$f(X) = \prod f_i(X)^{m_i}.$$

Then each $f_i(X)$ generates a prime ideal in $k[X]$, and determines a variety U_i. The intersection of S^n with H_f is the union of the U_i. Thus without loss of generality, we may assume that f is irreducible in $k[X]$. A generic point for U can then be constructed as follows. Suppose that X_n actually appears in $f(X)$, so that we may write

$$f(X) = g_d X_n^d + g_{d-1} X_n^{d-1} + \ldots + g_0$$

where g_i are polynomials in $k[X_1, \ldots, X_{n-1}]$. Let $x_1, \ldots, x_{n-1}$ be $n - 1$ algebraically independent quantities over k, and let x_n be a root of $f(X)$ viewed as a polynomial in X_n with coefficients in $k(x_1, \ldots, x_{n-1})$. Then $(x_1, \ldots, x_n)$ is a point in S^n of dimension $n - 1$, and $f(x) = 0$. Let U' be the variety having (x) as generic point over k. Then U' is obviously contained in U, and its dimension is $n - 1$. By Corollary 1 of Theorem 5, § 3, we must have $U = U'$, and the dimension theorem is therefore proved in case V is all of affine space.

The general case will now be reduced to the preceding one, and I am indebted to John Tate for the following elegant proof.

Let V^r be an arbitrary variety in S^n, with generic point (x) over k. Let $(y_1, \ldots, y_r)$ be elements of $k[x]$ algebraically independent over k, and such that $k[x]$ is integral over $k[y]$. Then (y) is a generic point of S^r over k. Let $f(X) = 0$ be the equation of our hypersurface, and let $z = f(x)$. We may assume that the intersection $V \cap H_f$ consists of one component W only. Indeed, if $W_1, \ldots, W_s$ are the distinct components of this intersection, we let $h_j(X)$ be polynomials vanishing on $W_j (j = 2, \ldots, s)$ and not on W_1. Then they do not vanish on V, since $W_1 \subset V$. We may now replace V by the variety V^* which is the locus of $(x, 1/h_2(x), \ldots, 1/h_s(x))$ over k. The intersection of V^* with $f(x) = 0$ consists of only one component, and it clearly suffices to prove that this component has dimension $r - 1$.

Let W be the unique component of $V \cap H_f$ and let (x') be a generic point of W over k. The homomorphism $k[x] \to k[x']$ induces a homomorphism $k[y] \to k[y']$. Let U be the locus of (y') over k. Since each coordinate y'_j is a polynomial function of (x'), we have $k(y') \subset k(x')$, and hence it will suffice to prove that U is a hypersurface section of S^r. This is done as follows. First note that the norm of $f(x)$ taken from $k(x)$ to $k(y)$ lies in the polynomial ring $k[y]$. Indeed, $f(x)$ (being an element of $k[x]$) is integral over $k[y]$, and so are all its conjugates. By Proposition 9 of § 6, so is the norm, i.e. the product of all the conjugates raised to a suitable power of the characteristic. Since $k[y]$ is integrally closed (because it is a unique factorization domain) the norm of $f(x)$ lies in $k[y]$. Denote it by $f_0(y)$. We shall prove that U is the intersection of S^r with the hypersurface $f_0(Y) = 0$. In the first place, since $f(x') = 0$, it is clear that $f_0(y') = 0$. Conversely, let $g(Y)$ be any polynomial in $k[Y]$ vanishing on U, i.e. such that $g(y') = 0$. Viewed as an element of $k[x]$, $g(y)$ must vanish on W, and hence by the Hilbert Nullstellensatz, we can write

$$g(y)^v = f(x)h(x)$$

with a suitable $h(x) \, \epsilon \, k[x]$. Taking the norm of both sides, we find

$$g(y)^m = f_0(y)h_0(y)$$

for a suitable power m. This proves that $g(y)$ vanishes on the hypersurface $f_0(Y) = 0$. We have thus proved that U is a hypersurface section of S^r, thereby concluding the proof of the theorem.

From the dimension theorem concerning the intersection of a variety with a hypersurface, we obtain the dimension theorem concerning the intersection of two varieties by a very simple trick.

COROLLARY. *Let V^r and W^s be two varieties in S^n. If $V \cap W$ is not empty, then every component of the intersection $V \cap W$ has dimension $\geq r + s - n$.*

Proof: Intersect the product $V \times W$ with the diagonal, i.e. with the hypersurfaces $X_1 - Y_1 = 0, \ldots, X_n - Y_n = 0$ and apply the dimension theorem already proved, together with the corollary to Theorem 10, § 6.

REMARK. Later, we shall define homogeneous varieties, which determine varieties in projective space. These homogeneous varieties will always have a point in common, namely the origin, and hence the intersection of two homogeneous varieties is never empty. This will show that if V and W are varieties in projective space of dimension n, and if $r + s - n$ is $\geqq 0$, then the intersection cannot be empty, and the dimension theorem applies.

8. Homogeneous varieties

Consider the polynomial ring $k[X]$. By a *monomial* $M(X)$ we shall always mean a monomial with coefficient 1, i.e. an expression of type $X_1^{\nu_1} \ldots X_n^{\nu_n}$, and $\sum \nu_j$ is called its degree. A polynomial $f(X)$ can then be written $f(X) = \sum c_\alpha M_\alpha(X)$ with c_α in k. If in this expression for f the degrees of the M_α are all the same, then we say that f is a *form*, or that it is *homogeneous* (of that degree). An arbitrary polynomial $f(X)$ in $k[X]$ can also be written $f(X) = \sum f^{(d)}(X)$ where each $f^{(d)}$ is a form of degree d (which may be 0), and $f^{(d)}$ is called the *homogeneous part of f of degree d*.

An ideal $\mathfrak{a}$ of $k[X]$ is called *homogeneous* if each homogeneous part $f^{(d)}$ of a polynomial f in $\mathfrak{a}$ is also in $\mathfrak{a}$.

PROPOSITION 1. *An ideal $\mathfrak{a}$ is homogeneous if and only if $\mathfrak{a}$ has a basis consisting of forms.*

Proof: If $\mathfrak{a}$ is homogeneous, and f_j are the elements of a basis, then the homogeneous parts of the f_j constitute also a basis. Conversely, let f be a form in $\mathfrak{a}$. Let g be arbitrary in $k[X]$. Write $g = \sum g^{(d)}$ where $g^{(d)}$ is the homogeneous part of g of degree d. Then $gf = \sum g^{(d)}f$ and each $g^{(d)}f$ is in $\mathfrak{a}$. Letting f range over a basis of $\mathfrak{a}$, we see that $\mathfrak{a}$ is homogeneous.

If (x) is a point in S^n, and t a quantity of Ω, we denote by (tx) the point $(tx_1, \ldots, tx_n)$. An algebraic set A is called *homogeneous* if for every point (x) in A, the points (tx) are also in A, for all t in Ω.

PROPOSITION 2. *An algebraic set A is homogeneous if and only if its associated ideal $\mathfrak{a}$ in $k[X]$ is homogeneous.*

Proof: Suppose $\mathfrak{a}$ is homogeneous. Let (x) be in A. If f is in $\mathfrak{a}$ and is homogeneous, then $f(x) = 0$, and hence $f(tx) = 0$ also. Since $\mathfrak{a}$ has a basis consisting of forms, this proves that (tx) is a zero of $\mathfrak{a}$ and hence is also in A. Hence A is homogeneous. Conversely, let A be homogeneous. Suppose f is in $\mathfrak{a}$, and write $f = \sum f^{(d)}$. Let (x) be in A. Select t transcendental over $k(x)$. Then we get $0 = f(x) = \sum t^d f^{(d)}(x)$, and hence each $f^{(d)}(x) = 0$. This is true for any point (x) in A, and consequently each $f^{(d)}$ must lie in $\mathfrak{a}$, and $\mathfrak{a}$ is homogeneous.

Let V be a homogeneous variety. Then we have just seen that its associated prime ideal $\mathfrak{p}$ in $k[X]$ is homogeneous. Let (x) be a generic point of V. We know that (tx) is in V for all $t \, \epsilon \, \Omega$. Hence $(x) \rightarrow (tx)$ is a specialization for t transcendental over $k(x)$. It is also obvious that (x) is a specialization of (tx).

We define a point (x) to be *homogeneous* if $(x) \rightarrow (tx)$ is a specialization. The generic point of a homogeneous variety is homogeneous, and conversely it is clear that if (x) is homogeneous then its associated prime ideal in $k[X]$, and the variety V which is its locus over k, are homogeneous.

PROPOSITION 3. *Let A be a homogeneous algebraic set. Then each component variety V of A is also homogeneous.*

Proof: Given a component V of A, let (x) be a generic point of V. We know $(tx) \, \epsilon \, A$. Since $(tx) \rightarrow (x)$ is a specialization for t transcendental over $k(x)$, the locus of (tx) is a variety containing V, and contained in A. Since we assumed V to be a component it must be V, and hence (tx) is in V, which is homogeneous.

To get projective varieties from homogeneous varieties, we exclude the point (0) and identify (x) with (tx) for all $t \, \epsilon \, \Omega$. However, in what follows it will be more convenient to deal with homogeneous varieties. We conclude by giving some indications on how to go from affine to projective varieties.

With each polynomial $f \, \epsilon \, k[X_1, \ldots, X_n]$ we associate a polynomial f^* in the polynomial ring $k[X_0, \ldots, X_n]$ obtained by adding one more variable X_0, with f^* defined by

$$f^* = X_0^d f(X_1/X_0, \ldots, X_n/X_0),$$

where d is the degree of f. If $d = \deg f$ and $e = \deg g$, then $d + e = \deg fg$. Hence $(fg)^* = f^*g^*$. Furthermore, f^* is a form, as one sees immediately from the fact that $f^*(tX_0, tX) = t^d f^*(X_0, X)$.

To each form $F(X_0, X)$ in $k[X_0, X]$ we associate the polynomial $F'(X) = F(1, X)$. We obviously have $(f^*)' = f$, but it is not always true that $(F')^* = F$.

Let $\mathfrak{a}$ be an ideal of $k[X]$. We let $\mathfrak{a}^*$ be the ideal in $k[X_0, X]$ generated by all forms $F \in k[X_0, X]$ such that F' lies in $\mathfrak{a}$. It follows that $\mathfrak{a}^*$ is a homogeneous ideal.

If (x) is a zero of $\mathfrak{a}$, and $t \in \Omega$, then (t, tx) is a zero of $\mathfrak{a}^*$. Indeed, if $F(X_0, X)$ is a form such that $F(1, X) \in \mathfrak{a}$, then $F(t, tx) = t^d F(1, x) = 0$. This implies that a basis of $\mathfrak{a}^*$, and hence $\mathfrak{a}^*$, vanishes on (t, tx).

PROPOSITION 4. *Let $\mathfrak{p}$ be a prime ideal in $k[X]$ with generic zero (x). Then $\mathfrak{p}^*$ is a prime ideal in $k[X_0, X]$ with generic zero (t, tx), where t is transcendental over $k(x)$.*

Proof: We have just seen that (t, tx) is a zero of $\mathfrak{p}^*$. Let $\mathfrak{q}^*$ be the prime ideal of $k[X_0, X]$ vanishing on (t, tx). Then $\mathfrak{v}^*$ is homogeneous and contains $\mathfrak{p}^*$. If F is a form in $\mathfrak{q}^*$ then $F(t, tx) = 0$ implies $t^d F(1, x) = 0$, and hence $F(1, x) = 0$. Hence $F' \in \mathfrak{p}$, and by definition, $F \in \mathfrak{p}^*$. This proves $\mathfrak{p}^* = \mathfrak{q}^*$, as desired.

Let V be the variety of zeros of $\mathfrak{p}$, and V^* the variety of zeros of $\mathfrak{p}^*$ (of course V^* is in S^{n+1}). If a point (x_0, x) in V^* has first component $x_0 \neq 0$, we may rewrite this point $(x_0, x_0 y)$ where $(y) = (x/x_0)$. It is then clear that (y) is a zero of $\mathfrak{p}$.

We note finally that the dimension of V^* is equal to the dimension of V plus 1.

9. Elimination theory

Throughout this section we work with homogeneous objects, in S^n.

The problem of elimination theory consists in giving algebraic criteria for certain algebraic sets to be non-empty. More precisely, a homogeneous ideal always has a zero, namely the origin (0), which will be called the *trivial zero*. We shall want to know when a homogeneous algebraic set has a non-trivial zero.

THEOREM 12. (*Hilbert-Zariski*). *Let V be a homogeneous variety with generic point (x). Let A be the algebraic set of zeros of the homogeneous ideal generated by forms $f_1, \ldots, f_s$ in $k[X]$. Then $V \cap A$ has only the trivial zero if and only if each x_i is integral over the ring $k[f(x)] = k[f_1(x), \ldots, f_s(x)]$.*

Proof: Suppose that some x_i is not integral over the ring $k[f(x)]$. We shall construct a non-trivial zero of $V \cap A$. There exists a place φ finite on this ring but not finite on some x_i. Say x_n has maximal value under this place (i.e. under the valuation associated with the place we have $|x_n| \geq |x_i|$ for all i), and let $y_i = x_i/x_n$. Then $y_n = 1$, and $\varphi(y_n) = 1$. Furthermore, φ is finite on k, hence an isomorphism on k. We may assume that φ is identity on k. We have

$$0 = \varphi(f_j(x)/x_n^{d_j}) = \varphi(f_j(y)) = f_j(y'),$$

where we put $y_i' = \varphi(y_i)$. Since (y') is obviously a specialization of (x), we have our zero.

(Of course we have constructed a zero which may not be algebraic over k. If one wants one which is algebraic over k, then one can use the Hilbert Nullstellensatz to get it).

Conversely, suppose that all x_i are integral over $k[f(x)]$. We can write

$$x_i^m + a_{m-1}(f)x_i^{m-1} + \ldots + a_0(f) = 0,$$

where the coefficients $a(f)$ are polynomials in the $f_j(x)$ with coefficients in k. The ideal $\mathfrak{p}$ in $k[X]$ vanishing on (x) is homogeneous. Since we have

$$X_i^m + a_{m-1}(f(X))X_i^{m-1} + \ldots + a_0(f(X)) \equiv 0 \pmod{\mathfrak{p}}$$

we can express this as an identity among the homogeneous components of each polynomial of degree m, and we see that an equation similar to the above holds with coefficients which do not have constant terms. Suppose that there is a non-trivial zero (y)

in $V \cap A$. Say $y_1 \neq 0$. Considering the equation for X_1, and substituting the value (y) for (X) we obtain

$$y_1^{m_1} = 0,$$

a contradiction.

COROLLARY. *Let $f_1, \ldots, f_s$ be forms in n variables, and suppose $n > s$. Then these forms have a non-trivial common zero.*

Proof: Let $V = S^n$ so that (x) consists of n algebraically independent quantitites over k. The transcendence degree of $k(x)$ over k is greater than that of $k(f(x))$ over k. Hence the x's cannot be algebraic over $k(f(x))$, let alone integral over the ring $k[f(x)]$. Hence a common zero exists.

Note that the corollary is also a consequence of the dimension theorem. That is, we know that there is a trivial common zero, and by the dimension theorem, every component of the intersection has dimension $\geq n - s \geq 1$.

We consider the product space $S^m \times S^n$ with the variables $(W) = (W_1, \ldots, W_m)$ and $(X) = (X_1, \ldots, X_n)$. Let (x) be a point of S^n and (w) a point in S^m. Then we may look at (w, x) as a point in S^{m+n}.

THEOREM 13. *Let V be a variety in S^{m+n} with generic point (w, x) which is homogeneous in (x), i.e. such that $(w, x) \to (w, tx)$ is a specialization for all quantities t, and $(x) \neq (0)$. Then every specialization $(w) \to (w')$ can be extended to a specialization $(w, x) \to (w', x')$ where $(x') \neq (0)$ (i.e. (x') is non-trivial).*

Proof: The homomorphism $k[w] \to k[w']$ can be extended to a place φ of $k(w, x)$. Say $|x_n| \geq |x_i|$ for all i, under the associated valuation. Let $\xi_i = x_i/x_n$. Then all ξ_i are finite under the place. Put $(\xi') = \varphi(\xi)$. Then all ξ_i' are finite, and $\xi_n' = 1$. This is what we wanted, because $(w, x) \to (w', \xi')$ is a specialization.

We note parenthetically that the preceding result may be false if (w, x) is not homogeneous in (x). For instance, the variety defined by the equation $W_1^2 + W_2 X_1 = 0$ is a counterexample.

To apply Theorem 13 we make some definitions. By a set of forms $\mathscr{F}$ we shall mean a finite set of forms $(f) = (f_1, \ldots, f_r)$

with coefficients in Ω. We let $d_1, \ldots, d_r$ be their respective degrees. Each f_i can be written

$$f_i = \sum_\alpha w_{i,\alpha} M_\alpha(X),$$

where $M_\alpha(X)$ is a monomial in (X) and $w_{i,\alpha}$ is its coefficient.

We shall say that $\mathscr{F}$ has a non-trivial zero (x) if $(x) \neq (0)$ and $f_i(x) = 0$ for all $f \in \mathscr{F}$.

We let (w) be the point obtained by arranging the coefficients $w_{i,\alpha}$ of the forms in some definite order, and consider this point as a generic point over k. In other words, for given degrees $d_1, \ldots, d_r$ the set of all forms $\mathscr{F} = (f_1, \ldots, f_r)$ with these degrees may be viewed as being in $1 - 1$ correspondence with the points of a certain space S^m. It is the purpose of elimination theory to prove that those sets of forms $\mathscr{F}$ which have a non-trivial zero form an algebraic set in S^m.

Let $\mathscr{F}' = (f_1', \ldots, f_r')$ be another set of forms of the same degrees $d_1, \ldots, d_r$. Write

$$f_i' = \sum_\alpha w_{i,\alpha}' M_\alpha(X).$$

Then $\mathscr{F}'$ is another point in S^m, and we say that $\mathscr{F} \to \mathscr{F}'$ is a specialization if $(w) \to (w')$ is a specialization.

THEOREM 14. *Let $\mathscr{F}$ be a set of forms, viewed as a point in S^m, and let $\mathscr{F}'$ be a specialization of $\mathscr{F}$. If $\mathscr{F}$ has a non-trivial zero, then so does $\mathscr{F}'$.*

Proof: The argument is entirely similar to that of Theorem 13.

COROLLARY. *Let $\mathscr{F}$ be a set of n forms in n variables, and assume that $\mathscr{F}$ is a generic point of S^m, i.e. that the coefficients of these forms are algebraically independent over k. Then $\mathscr{F}$ does not have a non-trivial zero.*

Proof: There exists a specialization of $\mathscr{F}$ which has only the trivial zero, namely $f_1' = X_1^{d_1}, \ldots, f_n' = X_n^{d_n}$.

Let now $(F) = (F_1, \ldots, F_r)$ be a set of forms of degrees d_i in $k[W, X]$ of type

$$F_i(W, X) = \sum_\alpha W_{i,\alpha} M_\alpha(X),$$

where M_a ranges again over all monomials in (X) of degree d_i. Note that this time the coefficients (W) are themselves variables, and not quantities (i.e. elements of Ω).

Let $\mathfrak{a}$ be the ideal generated by these forms in the ring $k[W, X]$. The zeros (w, x) of $\mathfrak{a}$ form an algebraic set A in the product space, and A is homogeneous in (X). This algebraic set is a union of varieties, each of which is homogeneous in (X). It is then *a priori* clear from Theorem 13 that the projection of the components of A is an algebraic set. This is the well-known statement of elimination theory according to which the existence of non-trivial zeros for a set of forms is given by algebraic conditions on their coefficients. We note that the original forms are also homogeneous in (W), and hence so is the projection.

Following Van der Waerden, we can actually get a considerably stronger result: The above-mentioned algebraic set of zeros of $\mathfrak{a}$ is actually a variety, and we shall construct explicitly its generic point. We shall see that $\mathfrak{a}$ is not quite the prime ideal associated with this variety, but that it is very close to being this prime ideal.

We work in the ring $k[W, X]$. Suppose that a relation

(1) $$X_j^s \, G\,(W,X) \equiv 0 \qquad (\mathrm{mod}\ F_1, \ldots, F_r)$$

holds for some polynomial G in $k[W, X]$ and some power X_j^s of one of the variables X_j. We select any one of the variables, say X_n, and rewrite the forms F_i as follows:

$$F_i = F_i^* + Y_i \, X_n^{d_i}$$

where F_i^* is the sum of all monomials except that containing $X_n^{d_i}$. The coefficients (W) are thereby split into two groups, which we denote by (Y) and (Z). (Z) are the coefficients of F_i^*. We have $(W) = (Y, Z)$, and we write

$$F_i(Y, Z, X) = F_i^*(Y, X) + Z_i X_n^{d_i}.$$

Corresponding to the variables (Y, X) we choose quantities (y, x) algebraically independent over k. Let $w_i = -\,F_i^*(y, x)/x_n^{d_i}$. We contend that the point (y, z, x) is a generic point for A.

From our construction it is immediately clear that $F_i(y, z, x) = 0$ for all i, and consequently $G(y, z, x) = 0$ according to (1).

Conversely, suppose that $G(y, z, x) = 0$ for some G in $k[Y, Z, X]$ From Taylor's formula for several variables we obtain

$$G(Y, Z, X) = G(Y, -F_i^*/X_n^{d_i} + Z_i + F_i^*/X_n^{d_i}, X)$$
$$= G(Y, -F_i^*/X_n^{d_i}, X) + \sum (Z_i + F_i^*/X_n^{d_i})^\mu H(Y, Z, X),$$

where the sum is taken over terms having one factor $(Z_i + F_i^*/X_n^{d_i})$ to some power $\mu > 0$, and some factor H in $k[Y, Z, X]$.

From the way in which (y, z, x) was constructed, and the fact that $G(y, z, x) = 0$, we see that one term vanishes, and hence

$$G(Y, Z, X) = \sum (Z_i + F_i^*/X_n^{d_i})^\mu H(Y, Z, X).$$

Clearing denominators of X_n, we get

$$X_n^s G(Y, Z, X) \equiv 0 \qquad (\text{mod } F_1, \ldots, F_r)$$

or in the original notation,

(2) $\qquad X_n^s G(W, X) \equiv 0 \qquad (\text{mod } F_1, \ldots, F_r)$

for some integer s.

We have therefore proved the following statements: Given G in $k[W, X] = k[Y, Z, X]$. If (1) holds for some j, then (1) holds for every j. Furthermore, if $G(z, x) = 0$ then (2) holds and consequently (1) holds for every j.

Let $\mathfrak{p}$ be the ideal in $k[W, X]$ vanishing on the point (w, x). Then $\mathfrak{p}$ is prime. From (1) we see that every zero of $\mathfrak{a}$ is a zero of $\mathfrak{p}$. But the generic zero of $\mathfrak{p}$ is a zero of $\mathfrak{a}$. Hence A is the variety associated with $\mathfrak{p}$. Relations (1) show how close $\mathfrak{a}$ is to being the prime ideal associated with this variety. If $\mathfrak{p}_0$ is the ideal $\mathfrak{p} \cap k[W]$ then $\mathfrak{p}_0$ is prime. According to Theorem 13, the set theoretic projection V_0 of our variety on S^m is a variety, and (w) is a generic point of V_0, which is the variety associated with $\mathfrak{p}_0$.

We leave the proof of the next theorem as an exercise to the reader.

THEOREM 15. *For $n-1$ forms in n variables, the variety of coefficients for which the forms have a non-trivial zero is the entire space. For n forms in n variables, it is a hypersurface, i.e. it consists of the zeros of a single polynomial $G(W)$.*

CHAPTER III

Absolute Theory of Varieties

The theory presented in this chapter is at first intended to deal with the problem of constant field extensions: What happens to a k-variety V when we extend the ground field k? It does not always remain a variety, since one could for instance start with a pair of conjugate points $(1, \sqrt{2})$ and $(1, -\sqrt{2})$ over the rational numbers **Q**. These form a **Q**-variety, but become two distinct varieties over the reals, namely two distinct points. Apart from this kind of behavior, the characteristic contributes other troubles. It turns out that one can formulate an elegant necessary and sufficient condition on a k-variety in order to insure that it behaves well under all extensions of the ground field, namely the condition of regularity. It is also this condition which guarantees that a product of varieties remains a variety (over the given field k).

The reader should not have the impression, however, that this chapter is used exclusively as a technical means to regularize the behavior of varieties under constant field extensions. In fact, it yields the foundations for a theory of algebraic systems of varieties. Even if we decided from the start to consider only varieties defined over an algebraically closed field, we would soon be lead to abandon this decision. For example, let k be algebraically closed. Let (x) be the generic point of a k-variety V over k, and consider an intermediate field between k and $k(x)$. It is easily shown that such a field is finitely generated, i.e. can be written $k(y)$, where (y) is the generic point of a k-variety W over k. Then of course $k(y)$ is not algebraically closed. The point (x) can be viewed as the generic point of a $k(y)$-variety U, which is the locus of (x) over $k(y)$. This variety U is viewed as depending on the parameters (y), i.e. viewed as a "generic" variety in an algebraic system which gives a sort of fibering of the variety V.

Much information concerning V can be obtained by knowing the two steps in the tower

$$k \subset k(y) \subset k(x).$$

The greater part of this chapter stems from the first chapter of *Foundations*. It is the only chapter in this book which is of a strictly technical algebraic nature, and we have tried to make all the subsequent chapters independent of this one, except for isolated sections dealing specifically with problems of constant field extension. Naturally, the contents of this chapter form an indispensable basis for all the purely algebraic methods of algebraic geometry.

1. Auxiliary algebraic results

In this section we discuss three types of extensions of a field: separable, regular, and primary.

We shall assume known the theory of finite algebraic extensions. Among the facts used most frequently, it is worth while to recall the following ones. Let K be a finite algebraic extension of k, of degree m. There is a maximal subfield E of K containing k and separable over k. K is purely inseparable over E. We denote the degree $[K : E]$ by $[K : k]_i$ and $[E : k]$ by $[K : k]_s$. These are called the inseparable and separable degrees of K over k. If K is separable over k (resp. purely inseparable) and L is an arbitrary extension of k then KL is separable (resp. purely inseparable) over L. We call the extension KL over L the *translation of K over L*.

In addition to the above, we also recall some terminology of vector spaces. Let E be a vector space over a field K. A set of elements $\{u_a\}$ of E is said to generate E over K if every element v of E can expressed as a linear combination $v = \sum c_a u_a$ with $c_a \, \epsilon \, K$, and almost all $c_a = 0$. ("Almost all" means all but a finite number.) If the elements $\{u_a\}$ are linearly independent over K then they are called a basis of E over K. Using Zorn's lemma, one shows that every vector space has a basis, and that a basis can always be selected from a given set of generators.

a. *Linear disjointness and algebraic independence*

We shall discuss the way in which two extensions K and L of a field k behave with respect to each other.

K is said to be *linearly disjoint from L over k* if every finite set of elements of K that is linearly independent over k is still such over L.

The definition is unsymmetric, but we prove right away that the property of being linearly disjoint is actually symmetric for K and L. Assume K linearly disjoint from L over k. Let $y_1, \ldots, y_n$ be elements of L linearly independent over k. Suppose there is a non-trivial relation of linear dependence over K,

$$(1) \qquad x_1 y_1 + x_2 y_2 + \ldots + x_n y_n = 0.$$

Say $x_1, \ldots, x_r$ are linearly independent over k, and $x_{r+1}, \ldots, x_n$ are linear combinations $x_i = \sum_{\mu=1}^{r} a_{i\mu} x_\mu$, $i = r+1, \ldots, n$. We can write the relation (1) as follows:

$$\sum_{\mu=1}^{r} x_\mu y_\mu + \sum_{i=r+1}^{n} (\sum a_{i\mu} x_\mu) y_i = 0$$

and collecting terms, after inverting the second sum, we get

$$\sum_{\mu=1}^{r} (y_\mu + \sum_{i=r+1}^{n} (a_{i\mu} y_i)) x_\mu = 0.$$

The y's are linearly independent over k, so the coefficients of x_μ are $\neq 0$. This contradicts the linear disjointness of K and L over k.

We should remark that we could have defined K and L to be linearly disjoint over k by saying that that canonical map

$$K \otimes L \to K[L]$$

of the tensor product (taken over k) onto the vector space generated over k by all products xy with $x \in K$ and $y \in L$ is a k-isomorphism. The equivalence between this condition and the unsymmetric one we have taken is easily proved, and properties of linearly disjoint fields can then be deduced from properties of tensor products if the reader is acquainted with this notion. For instance, Proposition 1 below expresses no more than the formula

$$K \otimes_k L = (K \otimes_k E) \otimes_E L.$$

However, for the convenience of the reader not acquainted with the tensor product, we reproduce the proofs.

We now give two criteria for linear disjointness.

Criterion 1. Suppose that K is the quotient field of a ring R and L the quotient field of a ring S. To test whether L and K are linearly disjoint, it suffices to show that if elements $y_1, \ldots, y_n$ of S are linearly independent over k, then there is no linear relation among the y's with coefficients in R. Indeed, if elements $y_1, \ldots, y_n$ of L are linearly independent over k, and if there is a relation $x_1 y_1 + \ldots + x_n y_n = 0$ with $x_i \in K$, then we can select y in S and x in R such that $xy \neq 0$, $yy_i \in S$ for all i, and $xx_i \in R$ for all i. Multiplying the relation by xy gives a linear dependence between elements of R and S. However, the yy_i are obviously linearly independent over k, and this proves our criterion.

Criterion 2. Again let R be a subring of K such that K is its quotient field and R is a vector space over k. Let $\{u_\alpha\}$ be a basis of R considered as a vector space over k. To prove K and L linearly disjoint over k, it suffices to show that the elements $\{u_\alpha\}$ of this basis remain linearly independent over L. Indeed, suppose this is the case. Let $x_1, \ldots, x_m$ be elements of R linearly independent over k. They lie in a finite dimension vector space generated by some of the u_α, say $u_1, \ldots, u_n$. They can be completed to a basis of this space over k. Lifting this vector space of dimension n over L, it must conserve its dimension because the u's remain linearly independent by hypothesis, and hence the x's must also remain linearly independent.

The next proposition gives a useful criterion which allows us to recognize linear disjointness in a tower of fields.

PROPOSITION 1. *Let K be field containing another field k, and let $L \supset E$ be two other extensions of k. Then K and L are linearly disjoint over k if and only if K and E are linearly disjoint over k and KE, L are linearly disjoint over E.*

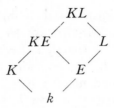

Proof: Assume first that K, E are linearly disjoint over k, and KE, L are linearly disjoint over E. Let $\{\varkappa\}$ be a basis of K as vector space over k (we use the elements of this basis as their own indexing set), and let $\{\alpha\}$ be a basis of E over k. Let $\{\lambda\}$ be a basis of L over E. Then $\{\alpha\lambda\}$ is a basis of L over k. If K and L are not linearly disjoint over k, then there exists a relation

$$\sum_{\lambda,\,\alpha} (\sum_{\varkappa} c_{\varkappa\lambda a}\,\varkappa)\lambda\alpha = 0 \qquad \text{with some } c_{\varkappa\lambda a} \neq 0.$$
$$c_{\varkappa\lambda a} \in k.$$

Changing the order of summation gives

$$\sum_{\lambda} (\sum_{\varkappa,\,\lambda} c_{\varkappa\lambda a}\,\varkappa\alpha)\lambda = 0$$

contradicting the linear disjointness of L and KE over E.

Conversely, assume that K and L are linearly disjoint over k. Then *a fortiori*, K and E are also linearly disjoint over k, and the field KE is the quotient field of the ring $E[K]$ generated over E by all elements of K. This ring is a vector space over E, and a basis for K over k is also a basis for this ring $E[K]$ over E. With this remark, and the criteria for linear disjointness, we see that it suffices to prove that the elements of such a basis remain linearly independent over L. At this point we see that the arguments given in the first part of the proof are reversible. We leave the formalism to the reader.

We introduce another notion concerning two extensions K and L of a field k. We shall say that K is *free from L over k* if every finite set of elements of K algebraically independent over k remains such over L. If (x) and (y) are two sets of quantities, we say that they are *free over k* (or *independent over k*) if $k(x)$ and $k(y)$ are free over k.

Just as with linear disjointness, our definition is unsymmetric, and we prove that the relationship expressed therein is actually symmetric. Assume therefore that K is free from L over k. Let $y_1, \ldots, y_n$ be elements of L, algebraically independent over k. Suppose they become dependent over K. They become so in a subfield F of K finitely generated over k, say of transcendence degree r over k. Computing the transcendence degree of $F(y)$ over k in two ways gives a contradiction:

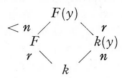

PROPOSITION 2. *If K and L are linearly disjoint over k then they are free over k.*

Proof: Let $x_1, \ldots, x_n$ be elements of K algebraically independent over k. Suppose they become algebraically dependent over L. We get a relation

$$\sum y_\alpha M_\alpha(x) = 0$$

between monomials $M_\alpha(x)$ with coefficients y_α in L. This gives a linear relation among the $M_\alpha(x)$. But these are linearly independent over k because the x's are assumed algebraically independent over k. This is a contradiction.

PROPOSITION 3. *Let L be an extension of k, and let $(u) = (u_1, \ldots, u_r)$ be a set of quantities algebraically independent over L. Then the field $k(u)$ is linearly disjoint from L over k.*

Proof: According to the criteria for linear disjointness, it suffices to prove that the elements of a basis for the ring $k[u]$ that are linearly independent over k remain so over L. In fact the monomials $M(u)$ give a basis of $k[u]$ over k. They must remain linearly independent over L, because as we have seen, a linear relation gives an algebraic relation. This proves our proposition.

Note finally that the property that two extensions K and L of a field k are linearly disjoint or free is of finite type. To prove that they have either property, it suffices to do it for all subfields K_0 and L_0 of K and L respectively which are finitely generated over k. This comes from the fact that the definitions involve only a finite number of quantities at a time.

b. *Separable extensions*

Let K be a finitely generated extension of k, $K = k(x)$. We shall say that it is *separably generated* if we can find a transcendence basis $(t_1, \ldots, t_r)$ of K/k such that K is separably algebraic over $k(t)$. Such a transcendence base is said to be a *separating transcendence base* for K over k.

We always denote by p the characteristic if it is not 0. The field obtained from k by adjoining all p^mth roots of all elements of k will be denoted by k^{1/p^m}. The compositum of all such fields for $m = 1, 2, \ldots$, is denoted by k^{1/p^∞}.

THEOREM 1. *The following conditions concerning an extension field K of k are equivalent*:

(1) K *is linearly disjoint from* k^{1/p^∞}.

(2) K *is linearly disjoint from* k^{1/p^m} *for some* m.

(3) *Every subfield of K containing k and finitely generated over k is separably generated.*

Proof: It is obvious that (1) implies (2). In order to prove that (2) implies (3), we may clearly assume that K is finitely generated over k, say $K = k(x) = k(x_1, \ldots, x_n)$. Let the transcendence degree of this extension be r. If $r = n$, the proof is complete. Otherwise, say $x_1, \ldots, x_r$ is a transcendence base. Then x_{r+1} is algebraic over $k(x_1, \ldots, x_r)$. Let $f(X_1, \ldots, X_{r+1})$ be a polynomial of lowest degree such that $f(x_1, \ldots, x_{r+1}) = 0$. Then f is irreducible. We contend that not all $x_i(i = 1, \ldots, r + 1)$ appear to the pth power throughout. If they did, we could write $f(X) = \sum c_a M_a(X)^p$ where $M_a(X)$ are monomials in $X_1, \ldots, X_{r+1}$ and $c_a \in k$. This would imply that the $M_a(x)$ are linearly dependent over $k^{1/p}$ (taking the pth root of the equation $\sum c_a M_a(x)^p = 0$). However, the $M_a(x)$ are linearly dependent over k (otherwise we would get an equation for $x_1, \ldots, x_{r+1}$ of lower degree) and we thus get a contradiction to the linear disjointness of $k(x)$ and $k^{1/p}$. Say X_1 does not appear to the pth power throughout, but actually appears in $f(X)$. We know that $f(X)$ is irreducible in $k[X_1, \ldots, X_{r+1}]$ and hence $f(x) = 0$ is an irreducible equation for x_1 over $k(x_2, \ldots, x_{r+1})$. Since X_1 does not appear to the pth power throughout, this equation is a separable equation for x_1 over $k(x_2, \ldots, x_{r+1})$, in other words, x_1 is separable algebraic over $k(x_2, \ldots, x_{r+1})$. From this it follows that it is separable algebraic over $k(x_2, \ldots, x_n)$. If $(x_2, \ldots, x_n)$ is a transcendence base, the proof is complete. If not, say that x_2 is separable over $k(x_3, \ldots, x_n)$. Then $k(x)$ is separable over $k(x_3, \ldots, x_n)$. Proceeding inductively, we see that the procedure can be continued until we get down to a transcendence base. This proves that (2) implies (3). It also proves that a separating transcendence base for $k(x)$ over k can be selected from the given set of generators (x).

To prove that (3) implies (1) we may assume that K is finitely generated over k. Let (u) be a transcendence base for K over k. Then K is separably algebraic over $k(u)$. By Proposition 3, $k(u)$ and k^{1/p^∞} are linearly disjoint. Let $L = k^{1/p^\infty}$. Then $k(u)L$ is purely inseparable over $k(u)$, and hence is linearly disjoint from K over $k(u)$ by the elementary theory of finite algebraic extensions. Using Proposition 1, we conclude that K is linearly disjoint from L over k, thereby proving our theorem.

An extension K of k satisfying the conditions of Theorem 1 is called *separable*. This definition is compatible with the use of the word for algebraic extensions.

The first condition of our theorem is known as *MacLane's criterion*. It has the following immediate corollaries.

COROLLARY 1. *If K is separable over k, and E is a subfield of K containing k, then E is separable over k.*

COROLLARY 2. *Let E be a separable extension of k, and K a separable extension of E. Then K is a separable extension of k.*
Proof: Apply Proposition 1 and the definition of separability.

COROLLARY 3. *If k is perfect, every extension of k is separable.*

COROLLARY 4. *Let K be a separable extension of k, and free from an extension L of k. Then KL is a separable extension of L.*

Proof: An element of KL has an expression in terms of a finite number of elements of K and L. Hence any finitely generated subfield of KL containing L is contained in a composite field FL, where F is a subfield of K finitely generated over k. By Corollary 1, we may assume that K is finitely generated over k. Let (t) be a transcendence base of K over k, so K is separable algebraic over $k(t)$. By hypothesis, (t) is a transcendence base of KL over L, and since every element of K is separable algebraic over $k(t)$, it is also separable over $L(t)$. Hence KL is separably generated over L. This proves the corollary.

COROLLARY 5. *Let K and L be two separable extensions of k, free from each other over k. Then KL is separable over k.*

Proof: Use Corollaries 4 and 2.

COROLLARY 6. *Let K, L be two extensions of k, linearly disjoint over k. Then K is separable over k if and only if KL is separable over L.*

Proof: If K is not separable over k, it is not linearly disjoint from $k^{1/p}$ over k, and hence *a fortiori* it is not linearly disjoint from $Lk^{1/p}$ over k. By Proposition 1, this implies that KL is not linearly disjoint from $Lk^{1/p}$ over L, and hence that KL is not separable over L. The converse is a special case of Corollary 4, taking into account that linearly disjoint fields are free.

We conclude our discussion of separability with two results. The first one has already been proved in the first part of Theorem 1, but we state it here explicitly for future reference.

PROPOSITION 4. *If K is a separable extension of k, and is finitely generated, then a separating transcendence base can be selected from a given set of generators.*

To state the second result we denote by K^{p^m} the field obtained from K by raising all elements of K to the p^mth power.

PROPOSITION 5. *Let K be a finitely generated extension of a field k. If $K^{p^m}k = K$ for some m, then K is separably algebraic*

*over k. Conversely, if K is separably algebraic over k, then $K^{p^m}k = K$
for all m.*

Proof: If K/k is separably algebraic, then the conclusion follows
from the elementary theory of finite algebraic extensions. Conversely, if K/k is finite algebraic but not separable, then the
maximal separable extension of k in K cannot be all of K, and
hence $K^p k$ cannot be equal to K. Finally, if there exists an
element t of K transcendental over k, then $k(t^{1/p^m})$ has degree p^m
over $k(t)$, and hence there exists a t such that t^{1/p^m} does not lie
in K. This proves our proposition.

c. *Regular extensions*

As usual, we put a bar over a field to denote its algebraic
closure.

THEOREM 2. *The following two conditions concerning an extension
K of a field k are equivalent.*

(1) *k is algebraically closed in K (i.e. every element of K algebraic
over k lies in k), and K is separable over k.*

(2) *K is linearly disjoint from $\bar{k}$ over k.*

Proof: Assume condition (2). Then by the results obtained
before, we know that K must be separable over k. It is obvious
that k must be algebraically closed in K. Hence we see that (2)
implies (1).

To prove the converse, we need a lemma.

LEMMA. *Let k be algebraically closed in an extension K. Let x be
a quantity algebraic over k. Then k(x) and K are linearly disjoint
over k, and $[k(x) : k] = [K(x) : K]$.*

Proof: Let $f(X)$ be the irreducible polynomial for x over k.
Then $f(X)$ remains irreducible over K. Otherwise, its factors
would have coefficients algebraic over k, hence in k. Powers of x
form a basis of $k(x)$ over k, hence the same powers form a basis
of $K(x)$ over K. This proves the lemma.

To prove (2) from (1) we may assume without loss of generality
that K is finitely generated over k, and it suffices to prove that it is
linearly disjoint from an arbitrary finite algebraic extension
L of k.

If L is separable algebraic over k, then it can be generated by one element, and we can apply the lemma.

More generally, let E be the maximal separable subfield of L containing k. Using Proposition 1 we see that it suffices to prove that KE and L are linearly disjoint over E. Let (t) be a separating transcendence base for K over k. Then K is separably algebraic over $k(t)$. Furthermore, (t) is also a separating transcendence base for KE over E, and KE is separable algebraic over $E(t)$. Thus KE is separable over E, and by definition it is linearly disjoint from L over E because L is purely inseparable over E. This proves our theorem.

An extension of k satisfying the conditions of Theorem 2 is called *regular*.

COROLLARY 1. *Let K be a regular extension of k, and let E be a subfield of K containing k. Then E is regular over k.*

COROLLARY 2. *Let E be a regular extension of k, and K a regular extension of E. Then K is a regular extension of k.*

Proof: Apply Proposition 1 and the definition of regularity.

COROLLARY 3. *If k is algebraically closed, then every extension of k is regular.*

To state the analogue for regularity of Corollary 4 of Theorem 1 we must strengthen Theorem 2.

THEOREM 3. *Let K be a regular extension of k, let L be an arbitrary extension of k, and assume that K and L are free over k. Then K and L are linearly disjoint over k.*

Proof: (I am indebted to Artin for the following proof.) Without loss of generality we may assume that K is finitely generated over k. Let $x_1, \ldots, x_n$ be elements of K linearly independent over k, and suppose we have a relation of linear dependence $x_1 y_1 + \ldots + x_n y_n = 0$ with $y_i \in L$. Let φ be a k-valued place of L. Let (t) be a transcendence base of K over k. By hypothesis, the elements of (t) remain algebraically independent over L, and hence φ can be extended to a place of KL which is indentity on (t). This place must then be an isomorphism on $k(t)$, and hence on K, which is a finite algebraic extension of $k(t)$ (Proposition 2 of Chapter I,

§ 4). After a suitable isomorphism, we may take a place equivalent to φ which is identity on K. Say y_n has maximal value under the valuation of the place. Divide the relation $\sum x_i y_i = 0$ by y_n to get $\sum x_i (y_i / y_n) = 0$ and apply the place. This gives a linear relation among the x_i with coefficients in $\bar{k}$, contradicting the linear disjointness of K and $\bar{k}$. This proves our theorem.

COROLLARY 4. *Let K be a regular extension of k, free from an extension L of k over k. Then KL is a regular extension of L.*

Proof: From the hypothesis, we deduce that K is free from the algebraic closure $\bar{L}$ of L over k. By the theorem, K is linearly disjoint from $\bar{L}$ over k. By Proposition 1, KL is linearly disjoint from $\bar{L}$ over L, and hence KL is regular over L.

COROLLARY 5. *Let K and L be two regular extensions of k, free from each other over k. Then KL is a regular extension of k.*

Proof: Use Corollaries 4 and 2.

COROLLARY 6. *Let K, L be two extensions of k, linearly disjoint over k. Then K is regular over k if and only if KL is regular over L.*

Proof: The proof is obtained by replacing $k^{1/p}$ by $\bar{k}$ and "separable" by "regular" in the proof of the analogous corollary to Theorem 1.

There is an intermediate kind of extension which we take the opportunity of mentioning here, namely those extensions for which k is algebraically closed in K (but K is not necessarily separable, and hence regular over k). We shall prove for this kind of extension a result analogous to the above corollary. The reader will immediately see that this result can be used to give an alternative proof of Theorem 3.

THEOREM 4. *Let k be algebraically closed in an extension K, and let K be free from L over k. If L is separable over k, or K is separable over k, then L is algebraically closed in KL.*

Proof: Without loss of generality, we may assume that L is finitely generated, and then proceed inductively. If L is separable, then we can pick a separating transcendence base for L over k, and we are reduced to proving our theorem to the case where L is generated by an element transcendental over k, or separably algebraic over k. If K is separable, we are also reduced to proving the theorem to the case where L is generated by one element, transcendental over k, or algebraic (but not necessarily separably algebraic).

Suppose first that u is transcendental over K. If our theorem is false, there exist polynomials f, g in $K[u]$, say $f \notin k[u]$ and f, g relatively prime in $K[u]$, satisfying an equation

$$a_n(u)(f/g)^n + a_{n-1}(u)(f/g)^{n-1} + \ldots + a_0(u) = 0$$

with $a_i(u) \in k[u]$ and $a_0 \neq 0$. There exists an irreducible factor f_1 of f in $K[u]$ which is not in $k[u]$. Some root x of f_1 is not algebraic over k because the coefficients of f_1 are polynomial functions with integer coefficients of the roots. Hence $a_0(x) \neq 0$, $g(x) \neq 0$. Substituting x in the equation gives a contradiction.

Let us now deal with a quantity x algebraic over k. We assume that K is separable over k or that $k(x)$ is separable over k. Let y be in $K(x)$, and suppose y is algebraic over $k(x)$. We must show that y lies in $k(x)$. We contend that under our hypotheses, the field $k(x, y)$ can be generated by one element z over k. Indeed, if x is separable over k, then this is an elementary theorem (Van der Waerden, *Modern Algebra*, Section 40). Suppose K is separable over k. Let F be an intermediate field generated by a separating transcendence base of K over k. Then $K(x)$ is separable algebraic over $F(x)$, so y is separable over $F(x)$. But $k(x)$ is algebraically closed in the purely transcendental extension $F(x)$. Hence the equation for y over $F(x)$ is the same as over $k(x)$. Hence y is separable over $k(x)$. We can now use the elementary theorem again.

Using the lemma of Theorem 2, and the obvious fact that $K(x) = K(z)$, we have

$$[k(z) : k] = [K(z) : K] = [K(x) : K] = [k(x) : k].$$

Since $k(x) \subset k(z)$, we see that z and hence y lie in $k(x)$, as was to be shown.

$$
\begin{array}{ccccc}
& & K(x) & & \\
& \diagup & & \diagdown & \\
K(x) & & & & F(x,y) \\
& \diagdown & & \diagup & \\
& & F(x) & & \\
K & & & & k(x,y) \\
& \diagdown & & \diagup & \\
F & & & & k(x) \\
& & \diagdown & \diagup & \\
& & k & &
\end{array}
$$

d. *Primary extensions*

We shall denote by k_s the separable closure of a field k (i.e. the compositum of all separable algebraic extensions of k).

THEOREM 5. *The following two conditions concerning an extension K of k are equivalent*:

(1) *K is linearly disjoint from k_s over k.*

(2) *The algebraic closure of k in K is purely inseparable over k.*

Proof: It is obvious that (1) implies (2). Conversely, assuming (2), it suffices to prove that every finite separable algebraic extension of k is linearly disjoint from K over k. Such an extension can be generated by one element, and our theorem is an immediate consequence of the lemma of Theorem 2 together with Proposition 1.

If an extension K of k satisfies the conditions of Theorem 5 we shall say that it is *primary*. Using the second condition, we get

COROLLARY 1. *Let K be a primary extension of k, and let E be a subfield of K containing k. Then E is primary over k.*

COROLLARY 2. *Let E be a primary extension of k, and K a primary extension of E. Then K is a primary extension of k.*

Proof: Apply Proposition 1 and the definition of primary extensions.

COROLLARY 3. *If k is equal to k_s then every extension is primary.*

To state the analogue of the fourth corollary for primary extensions we must strengthen Theorem 5 in the same manner that we did for regular extensions.

THEOREM 6. *Let K be a primary extension of k, and L a separable extension of k, which is free from K over k. Then K and L are linearly disjoint.*

Proof: Without loss of generality, we may assume that K is finitely generated over k. Let (t) be a transcendence base of K over k, and put $F = k(t)$. For sufficiently high m, the field $K^{p^m}F$ is separable algebraic over F, and hence is a regular extension of k, which we denote by E. Note that K is purely inseparable over E. By hypothesis, E must be free from L, and hence linearly disjoint from L over k by Theorem 3. By Corollary 4 of Theorem 1, EL is separable over E, and by the definition of separability, EL is linearly disjoint from K over E. By Proposition 1 we conclude that K and L are linearly disjoint over k.

COROLLARY 4. *Let K be a primary extension of k. If L is a separable extension of k which is free from K over k, then KL is a primary extension of L.*

Proof: The separable closure L_s of L is separable over k by Corollary 2 of Theorem 1 and is free from K by hypothesis. By the theorem, it is linearly disjoint from K over k, and by Proposition 1 we see that KL is linearly disjoint from L_s over L. Hence KL is primary over L by definition.

COROLLARY 5. *Let K and L be two primary extensions of k free from each other over k. Then KL is a primary extension of k.*

Proof: Use Corollaries 4 and 2.

COROLLARY 6. *Let K, L be two extensions of k which are linearly disjoint over k. Then K is primary over k if and only if KL is primary over L.*

Proof: The proof is obtained by replacing $k^{1/p}$ by k_s and "separable" by "primary" in the proof of the analogous corollary to Theorem 1.

2. Behaviour of an ideal under constant field extensions

The pure theory of fields developed in Section 1 will now be tied up with the study of polynomial ideals and of varieties. As usual, if $(X) = (X_1, \ldots, X_n)$ is a set of variables, we mean by a monomial $M(X)$ a monomial with coefficient 1, $X_1^{v_1} \ldots X_n^{v_n}$.

We consider a field K and the polynomial ring $K[X]$ as a vector space over K. A basis for this space is given by all the monomials $M(X)$. If $\mathfrak{a}$ is an ideal of $K[X]$, then it is a subspace of $K[X]$, and we can consider $K[X]/\mathfrak{a}$ as a factor space over K. We see that the monomials generate this factor space, but are not necessarily a basis for it any more, i.e. are not linearly independent over K mod $\mathfrak{a}$.

Let $f \in K[X]$. We can write $f = \sum a_\alpha M_\alpha(X)$ with $a_\alpha \in K$ and all but a finite number of $a_\alpha = 0$. These a_α are called the coefficients of f. We say that an ideal $\mathfrak{a}$ has a *basis in k* if there exists a basis for $\mathfrak{a}$ consisting of polynomials f whose coefficients lie in k. If $\mathfrak{a}$ has a basis in k, we say that k is a *field of definition for* $\mathfrak{a}$, or that $\mathfrak{a}$ is *definable* in k. If σ is automorphism of K, we denote by f^σ the polynomial $\sum a_\alpha^\sigma M_\alpha(X)$, and by $\mathfrak{a}^\sigma$ the ideal consisting of all f^σ, where f ranges over $\mathfrak{a}$. The following theorem of Weil is fundamental.

THEOREM 7. *Let $\mathfrak{a}$ be an ideal of $K[X]$. There exists a minimal field of definition for $\mathfrak{a}$. This means, there exists a field $k_0 \subset K$ such that*

1. $\mathfrak{a}$ *has a basis in k_0.*
2. *If $\mathfrak{a}$ has a basis in k, then $k \supset k_0$.*

Furthermore, if σ is an automorphism of K then $\mathfrak{a}^\sigma = \mathfrak{a}$ if and only if σ leaves every element of k_0 fixed.

Proof: Let $B = \{M_\beta\}$ be a maximal set among all monomials M_α which are linearly independent over K mod $\mathfrak{a}$. In other words, B is a basis for $K[X]$ over K mod $\mathfrak{a}$, selected from among the monomials. Let $C = \{M_\gamma\}$ be the complement of B in the set of all monomials. For each M_γ in C we can write

$$M_\gamma \equiv \sum_\beta a_{\gamma,\beta} M_\beta \pmod{\mathfrak{a}}$$

uniquely with $a_{\gamma,\beta}$ in K. We have $M_\gamma - \sum a_{\gamma,\beta} M_\beta$ in $\mathfrak{a}$. Let k_0 be the field obtained by adjoining to the prime field all coefficients $a_{\gamma,\beta}$ for all γ, β. We contend that this is the field we are looking for.

We prove first that $\mathfrak{a}$ has a basis in k_0. Let $f \in \mathfrak{a}$. We can write

$$f = \sum_\gamma b_\gamma M_\gamma + \sum_\beta c_\beta M_\beta \qquad \text{with } b_\gamma, \ c_\beta \in K.$$

Correcting the coefficients of the M_β, we get

$$f = \sum_\gamma b_\gamma (M_\gamma - \sum_\beta a_{\gamma, \beta} M_\beta) + \sum_\beta d_\beta M_\beta$$

for some $d_\beta \in K$. But f and $M_\gamma - \sum a_{\gamma, \beta} M_\beta$ lie in $\mathfrak{a}$. This implies that $\sum d_\beta M_\beta \equiv 0 \pmod{\mathfrak{a}}$ and hence all $d_\beta = 0$ because the M_β are linearly independent mod $\mathfrak{a}$. This proves that our polynomials $M_\gamma - \sum a_{\gamma, \beta} M_\beta$ are a basis for $\mathfrak{a}$, which is therefore definable in k_0.

We shall now prove that if $\mathfrak{a}$ is definable in k, then k_0 is contained in k. Let $f_1, \ldots, f_r$ be a basis of $\mathfrak{a}$ in k. Let M_0 be any one of the M_γ. We can write

$$M_0 - \sum_\beta a_{0, \beta} M_\beta = g_1 f_1 + \ldots + g_r f_r$$

with $g_i \in K[X]$. We shall prove that the g_i can actually be taken in k. This will prove that the coefficients $a_{0, \beta}$ lie in k. This being true for every M_0 among the M_γ, we conclude that $k_0 \subset k$.

To each polynomial g_i, which can be written $\sum y_{i, \alpha} M_\alpha$ we associate a polynomial $G_i = \sum Y_{i, \alpha} M_\alpha$ where $Y_{i, \alpha}$ is an indeterminate if $y_{i, \alpha} \neq 0$, and $Y_{i, \alpha} = 0$ if $y_{i, \alpha} = 0$. In other words, G_i is like g_i except that it has generic coefficients. Then

$$G_1 f_1 + \ldots + G_r f_r = \sum_\gamma l_\gamma(Y) M_\gamma + \sum_\beta l_\beta(Y) M_\beta,$$

where the $l_\gamma(Y)$ and $l_\beta(Y)$ are linear forms in the (Y) with coefficients in k. The system of linear equations

$$l_0(Y) = 1$$

$$l_\gamma(Y) = 0$$

ranging for all $\gamma \neq 0$, and having coefficients in k, is solvable in K. Hence it is solvable in k. Let (y') be the special values solving this system in k, and let g_i' be the polynomials obtained by replacing (Y) by (y') in G_i. Then

$$g_1' f_1 + \dots + g_r' f_r = M_0 - \sum_\beta z_\beta M_\beta$$

for suitable coefficients z_β of M_β in k. The only M_γ appearing on the right-hand side is M_0.

The uniqueness of the expression for M_0 as a linear combination of M_β's mod $\mathfrak{a}$ now implies that the coefficients z_β are actually the coefficients $a_{0,\beta}$ which we had before. This proves the second part of our theorem.

Finally, if σ leaves k_0 fixed, then it leaves a basis for $\mathfrak{a}$ fixed, and hence maps $\mathfrak{a}$ onto itself. Conversely, suppose $\mathfrak{a}^\sigma = \mathfrak{a}$. Applying σ to the congruence defining M_γ, we get

$$M_\gamma \equiv \sum_\beta a_{\gamma,\beta}^\sigma M_\beta \qquad (\mathrm{mod}\ \mathfrak{a}).$$

Using the uniqueness of the coefficients $a_{\gamma,\beta}$, we see that they must be left fixed under σ. Since they generate k_0 over the prime field, it follows that k_0 is also fixed under σ, and our theorem is proved.

The arrangement of the above proof does not show *a priori* that k_0 is finitely generated over the prime field. However, we deduce this *a posteriori* from the following elementary result.

PROPOSITION 6. *Let K be a finitely generated extension of k, and let $K \supset E \supset k$. Then E/k is also finitely generated.*

Proof: E certainly has a finite transcendence base over k, which we may denote by $y_1, \dots, y_r$. Complete this base to a transcendence base $y_1, \dots, y_r, \dots, y_s$ of K over k. Then K is finitely generated over $k(y_1, \dots, y_s)$ and algebraic over it, hence finite algebraic. It suffices to prove that E is finite over $k(y_1, \dots, y_r)$. If this were not the case, E (as vector space over $k(y_1, \dots, y_r)$) would have infinite dimension. But by Proposition 3 of § 1, we know that E and $k(y)$ are linearly disjoint over $k(y_1, \dots, y_r)$ because E is algebraic over $k(y_1, \dots, y_r)$ and $y_{r+1}, \dots, y_s$ are independent over E. Translating E over $k(y)$, we get a contradiction because $E(y_1, \dots, y_s)$ is contained in K.

As an example, we shall determine the smallest field of definition for an ideal generated by linear polynomials:

$$l_i(X) = \sum c_{i\nu} X_\nu - a_i \qquad (i = 1, \ldots, m)$$

with $c_{i\nu}, a_i \in K$. We assume that the equations $l_i(X) = 0$, are consistent, i.e. that $\mathfrak{a}$ is not the unit ideal. After a suitable re-numbering of the l_i, we may assume that $l_1, \ldots, l_r$ is a maximal set of linearly independent polynomials among the l_i over K. The vector space of dimension $n + 1$ generated over K by 1, $X_1, \ldots, X_n$ is also generated by

$$1, l_1, \ldots, l_r, X_{r+1}, \ldots, X_n$$

after a suitable renumbering of the X_j. We can express $X_1, \ldots, X_r$ as linear combinations of these basis elements, so that

$$X_i = \sum_{\nu=r+1}^{n} d_{i\nu} X_\nu + e_i \ (\text{mod } \mathfrak{a}),$$

and we contend that the field k obtained by adjoining all $d_{i\nu}$, e_i to the prime field is the smallest field of definition for the ideal $\mathfrak{a}$. In fact, the linear forms $X_i - \sum_{\nu=r+1}^{n} d_{i\nu} X_\nu (i = 1, \ldots, r)$ must be linearly independent over K. Since $X_i - \sum d_{i\nu} X_\nu - e_i$ lie in $\mathfrak{a}$, they generate $\mathfrak{a}$. Hence $\mathfrak{a}$ is definable in k. On the other hand, the monomials $1, X_{r+1}, \ldots, X_n$ are linearly independent mod $\mathfrak{a}$. They can be included in a basis of $K[X]$ over K. From the way in which we defined the smallest field of definition of an ideal, we see that k is indeed this field.

Using similar arguments, one can determine the smallest field of definition of an ideal generated by one polynomial $f(X) = \sum c_\alpha M_\alpha(X)$. After dividing f by a constant, we may assume that one of the c_α is equal to 1, and we leave it as an exercise to the reader to prove that the smallest field of definition is the one obtained by adjoining to the prime field all the coefficients c_α.

In the rest of this section we consider prime ideals. Let (x) be a point, and let $\mathfrak{P}$ be the prime ideal determined by (x) in $K[X]$. If, as before, we let $B = \{M_\beta\}$ be a maximal set of monomials linearly independent mod $\mathfrak{P}$, then by definition, the quantities $M_\beta(x)$ form a linear base of the ring $K[x]$ viewed as vector space over K.

If k is a field contained in K, then the $M_\beta(X)$ are *a fortiori* linearly independent over k, because they are so over K, but they need not generate $k[x]$ over k.

THEOREM 8. *Let (x) be a point. Let K be a field, and k a subfield. Let $\mathfrak{P}$ be the ideal determined by (x) in $K[X]$. Then $\mathfrak{P}$ has a basis in k if and only if $k(x)$ and K are linearly disjoint over k.*

Proof: Suppose $k(x)$ and K are linearly disjoint over k.   Let $f(X)$ be in $\mathfrak{P}$. Then we can write $f(X) = \sum a_j f_j(X)$, where each $f_j(X)$ is in $k[X]$ and where the a_j are in K, and are linearly independent over k.   Now $f(x) = 0$ implies that $\sum a_j f_j(x) = 0$, and by linear disjointness, $f_j(x) = 0$ for all j. Hence each $f_j(X)$ is in $\mathfrak{P}$. It is also in $k[X]$. Hence the ideal generated by all $f_j(X)$ obtained in the above fashion from the polynomials of $\mathfrak{P}$ actually generates $\mathfrak{P}$, which has therefore a basis in k.

Conversely, assume that $\mathfrak{P}$ has a basis in k.   Let $\{M_j(X)\}$ $(j = 1, \ldots, r)$ be a finite set of monomials in (X) such that $\{M_j(x)\}(j = 1, \ldots, r)$ are linearly independent over k.   It will suffice to prove that the $M_j(x)$ remain linearly independent over K (using a criterion for linear disjointness, and the fact that the $\{M_\alpha(x)\}$ generate the vector space $k[x]$ over k). Suppose we have a linear relation $\sum a_j M_j(x) = 0$ with $a_j \, \epsilon \, K$.   Then $\sum a_j M_j(X)$ is in $\mathfrak{P}$ and we can write

$$\sum_{j=1}^{r} a_j \, M_j(X) = \sum_{i=1}^{s} b_i \, f_i(X)$$

where $f_i(X)$ is in $k[X]$, b_i is in K, and the $f_i(X)$ are linearly independent over k. The polynomials $M_1, \ldots, M_r, f_1, \ldots, f_s$ must be linearly independent over k, for otherwise we could write

$$\sum z_j M_j(X) + \sum y_i f_i(X) = 0$$

with $z_j, y_i \, \epsilon \, k$. Substituting (x) for (X) and using the hypothesis that the $M_j(x)$ are linearly independent over k, we see that $z_j = 0$ for all j, and hence $y_i = 0$. But being formal polynomials, they must remain linearly independent over K, so all $a_j = 0$; this proves the theorem.

COROLLARY 1. *Let (x) be a point, $\mathfrak{p}$ the prime ideal determined by (x) over k, and let K be an extension of k linearly disjoint from $k(x)$ over k. Then $\mathfrak{p}K[X]$ is the prime ideal $\mathfrak{P}$ determined by (x) over K.*

COROLLARY 2. *Let $\mathfrak{p}$ be a prime ideal in $k[X]$ and let K be a regular extension of k. Then $\mathfrak{p}K[X]$ is prime in $K[X]$.*

Proof: We can choose a generic zero (x) of $\mathfrak{p}$ such that $k(x)$ is free from K over k. It follows that $k(x)$ and K are linearly disjoint, and we can apply Corollary 1.

The preceding corollaries give us conditions under which a prime ideal remains prime under certain constant field extensions. We shall supplement them by stating another result which shows how one can come back down to $\mathfrak{p}$.

PROPOSITION 7. *Let $\mathfrak{a}$ be an ideal in $K[X]$. Suppose $\mathfrak{a}$ has a basis $f_1, \ldots, f_r$ in a subfield k of K, and let $\mathfrak{a}_0$ be the ideal in $k[X]$ generated by $f_1, \ldots, f_r$. Then $\mathfrak{a} \cap k[X] = \mathfrak{a}_0$.*

Proof: The proof is an immediate consequence of the following lemma.

LEMMA. *Let $\mathfrak{a}$ be an ideal in $k[X]$ and let K be an extension of k. Let f be in $\mathfrak{a}K[X]$, the ideal generated in $K[X]$ by $\mathfrak{a}$. Then f can be written $f = \sum c_j f_j(X)$ where $\{c_j\}$ is a set of elements of K linearly independent over k, and where $f_j(X) \in k[X]$. When f is so written, then all $f_j(X)$ lie in $\mathfrak{a}$. (They are in fact uniquely determined.)*

Proof: We can write $f = g_1 h_1 + \ldots + g_r h_r$ with h_ν in $\mathfrak{a}$ and each g_ν in $K[X]$. But g_ν is a sum of monomials $M(X)$ with coefficients in K. If M is a monomial, then Mh lies in $\mathfrak{a}$. If we express the coefficients in terms of a finite number of them, linearly independent over k, then we see that f has an expression $\sum c_j f_j(X)$ where $\{c_j\}$ are linearly independent over k, and the f_j are in $k[X]$ and even in $\mathfrak{a}$. However, since K and $k(X)$ are linearly disjoint over k (Proposition 3 of § 1) it follows that an expression as in the statement of the lemma determines the $f_j(X)$ uniquely. Since at least one expression is such that the f_j are in $\mathfrak{a}$, the proof is complete

In the following proposition, we shall give a description of what happens to a prime ideal under primary and separable extensions. If K is an extension of k, and $\mathfrak{p}$ is a prime ideal in $k[X]$, we denote by $\mathfrak{p}K[X]$ the ideal generated by $\mathfrak{p}$ in $K[X]$. Recall also that an ideal $\mathfrak{q}$ in $K[X]$ is said to be primary if $fg \in \mathfrak{q}$, $f \notin \mathfrak{q}$ implies that some power of g lies in $\mathfrak{q}$.

PROPOSITION 8. *Let $\mathfrak{p}$ be a prime ideal in $k[X]$, and let K be an extension of k. Let (x) be a generic zero of $\mathfrak{p}$ over k.*

(i) *If $k(x)$ is a primary extension of k, or if K is a primary extension of k, then $\mathfrak{p}K[X]$ is a primary ideal.*

(ii) *If $k(x)$ or K is a separable extension of k, then $\mathfrak{p}K[X]$ is the intersection of the prime ideals in $K[X]$ belonging to the algebraic set determined by $\mathfrak{p}K[X]$ over K.*

Proof: Without loss of generality, we may assume that K is finitely generated over k. After taking a suitable isomorphism of $k(x)$ over k, we may assume that $k(x)$ and K are free over k. We now proceed stepwise.

Let (t) be a transcendence base of K over k, and let $F = k(t)$. Then K is finite algebraic over F, and the fields FK^{p^m} form a decreasing sequence containing F. They must all become equal to some field E for large m, and by Proposition 5 of § 1, E is separable over F, and hence E is separable over k. If $k(x)$ is a primary extension of k, then $k(x)$ and E are linearly disjoint by Theorem 6 of § 1, and $\mathfrak{p}E[X]$ is prime by Corollary 2 of Theorem 8. If K is a primary extension of k, then E is regular over k (because it is separable), and again $k(x)$ is linearly disjoint from E over k so that $\mathfrak{p}E[X]$ is prime in $E[X]$. We are therefore reduced to proving (i) in case K is a purely inseparable extension of k. Let $\mathfrak{q} = \mathfrak{p}K[X]$. Suppose f, $g \in K[X]$, $fg \in \mathfrak{q}$ and no power of g is in $\mathfrak{q}$. For some integer n, K^{p^n} is contained in k, and hence g^{p^n} has coefficients in k. We can write $f(X)$ as a linear combination $\sum c_i f_i(X)$ where the c_i are elements of K linearly independent over k, and $f_i(X) \in k[X]$. It will suffice to prove that each f_i lies in $\mathfrak{p}$. We have

$$g^{p^n} f = \sum c_i (f_i g^{p^n})$$

and since $g^{p^n} f = g^{p^{n-1}}(gf)$ this expression lies in $\mathfrak{p}$. According to the lemma, we can conclude that $f_i g^{p^n}$ lies in $\mathfrak{p}$ for each i, and since $\mathfrak{p}$ is prime this implies that f_i lies in $\mathfrak{p}$, as was to be shown.

In order to prove (ii) assume first that K is separable over k. We may decompose K into a purely transcendental extension, followed by a separable algebraic one. Since the purely transcendental subfield of K is linearly disjoint from $k(x)$ over k, we are reduced to proving our result in the case K is finite separable algebraic over k. Let $f(X) \in K[X]$ vanish on the algebraic set determined by $\mathfrak{p}K[X]$ (or equivalently let $f(X)$ lie in the intersection of the prime ideals of $K[X]$ belonging to the K-varieties whose union constitutes the set of zeros of $\mathfrak{p}K[X]$). Let c_i be a basis of K over k, and write $f = \sum c_i f_i$ with $f_i \in k[X]$ (some of the f_i may be 0). Let σ range over all distinct isomorphisms of K over k. Then $f^\sigma = \sum c_i^\sigma f_i$ and the determinant $\det(c_i^\sigma)$ is not 0 because K is separable algebraic over k. We can therefore express each f_i as a linear combination

$$f_i = \sum d_{i\sigma} f^\sigma$$

with coefficients in the smallest Galois extension of k containing K. By the Hilbert Nullstellensatz, some power f^m of f lies in $\mathfrak{p}K[X]$. For each σ, $(f^\sigma)^m$ lies in $\mathfrak{p}K^\sigma[X]$, and hence $f^\sigma(x) = 0$. We see therefore that $f_i(x) = 0$, and since f_i lies in $k[X]$, this implies that $f_i \in \mathfrak{p}$. This concludes the proof in the case we are considering.

Assume finally that $k(x)$ is separable over k. We can find a subfield E of K containing k such that E is separable over k, and K purely inseparable over E. Applying the result we have just proved, we know that $\mathfrak{p}E[X]$ is an intersection of prime ideals, and one sees immediately that these prime ideals are such that their generic zeros over E generate a separable extension of E, which is linearly disjoint from K over E. To prove the last part of our proposition, it suffices therefore to make the following remark:

Let $\mathfrak{a}$, $\mathfrak{b}$ be two ideals in $k[X]$, and let K be an arbitrary extension of k. Then

$$(\mathfrak{a} \cap \mathfrak{b})K[X] = \mathfrak{a}K[X] \cap \mathfrak{b}K[X].$$

This is easily proved as follows. Let f be in the intersection on the right. We can write $f = \sum c_i f_i$ with $c_i \in K$ linearly independent over k, and $f_i \in k[X]$. Then f_i must lie in $\mathfrak{a} \cap \mathfrak{b}$, so f lies in the expression on the left. The converse is trivial, and Proposition 8 is completely proved.

3. Absolute varieties, fields of definition, generic points

We consider the ring $\Omega[X]$, where Ω is our universal domain. We still work in affine n-space $S^n = \Omega^n$.

By an *algebraic set* in S^n we mean the zeros of an ideal in $\Omega[X]$. A *variety* V (or better an affine variety) will be an algebraic set which is not the proper union of two algebraic sets. By the results of Chapter II, we have a decomposition of an algebraic set A into varieties, and this decomposition is unique if there is no inclusion relation among the varieties. An algebraic set is a variety if and only if its associated ideal $\mathfrak{P}$ in $\Omega[X]$ is a prime ideal. By the Hilbert Nullstellensatz, there is a $1 - 1$ correspondence between prime ideals in $\Omega[X]$ and varieties V, given by $V =$ zeros of $\mathfrak{P}$ and $\mathfrak{P} =$ ideal vanishing on V.

If $\mathfrak{P}$ is the prime ideal associated with a variety V in $\Omega[X]$, and if k is a field of definition for $\mathfrak{P}$, then we shall say that k is a field of *definition for V*. (*Warning*: this does not imply that if V is the set of zeros of an ideal with a basis in k then k is a field of definition for V. See below, the discussion concerning k-closed sets.) According to Theorem 7, every variety has a smallest field of definition, denoted by def (V). Let $k_0 = $ def (V). If k is a given field, then kk_0 is the smallest field of definition for V containing k.

Let k be a field of definition for V. Let $\mathfrak{p} = \mathfrak{P} \cap k[X]$. Let (x) be a generic zero of $\mathfrak{p}$ over k. We call (x) a *generic point of V over k*. Since $\mathfrak{p}$ and $\mathfrak{P}$ have a basis in common, it follows that V coincides with the k-variety of zeros of $\mathfrak{p}$ in the sense of Chapter II, and V is the locus of (x) over k.

Let $k \subset K$ be two fields of definition for V. Let $\mathfrak{p}_k$ and $\mathfrak{p}_K$ be the intersections of $\mathfrak{P}$ with $k[X]$ and $K[X]$ respectively. Let (x) be a generic point of V over K. By assumption, the ideal determined by (x) over K has a basis in k, and hence by Theorem 8 of § 2, $k(x)$ and K are linearly disjoint over k. Hence (x) is also a generic point of V over k. Furthermore $k(x)$ and K must be free over k (because they are linearly disjoint, Proposition 2 of § 1) and hence $\dim_k (x) = \dim_K (x)$. If k and k' are two fields of definition for V, then so is $kk' = K$. If (x) is a generic point of V over K, then it is a generic point of V over both k and k', and the dimension of (x) over K, k, or k' is the same. This proves that the dimension of a generic point over a field of definition is independent of the field of definition, and is called the *dimension of* V.

Finally we contend that if k is a field of definition for V and (x) is a generic point of V over k then $k(x)$ is a regular extension of k. Indeed, $\bar{k}$ is also a field of definition. Let $(\bar{x})$ be a generic point for V over $\bar{k}$. By the preceding arguments, $k(\bar{x})$ and $\bar{k}$ are linearly disjoint over k. Furthermore, (x) and $(\bar{x})$ are equivalent over k. Since $k(\bar{x})/k$ is regular, so is $k(x)/k$.

In the preceding discussion, we started with a prime ideal of $\Omega[X]$ and came down into the universal domain. We shall now show how to go up.

Let V be a k-variety in the sense of Chapter I. By definition, V consists of the zeros of a prime ideal $\mathfrak{p}$ in $k[X]$. We shall say that V is regular, and that $\mathfrak{p}$ is regular if a generic zero (x) of $\mathfrak{p}$ over k generates a regular extension of k. We found before that a k-variety obtained by coming down from $\Omega[X]$ is regular. Conversely, every k-variety which is regular gives rise to a variety over Ω. Indeed, if K is any field containing k, we can select a generic point (x) of V over k which is free from K. Then $k(x)$ and K are linearly disjoint over k by Theorem 3 of § 1. Hence by Theorem 8 of § 2 the locus of (x) over K is the same as the locus of (x) over k and $\mathfrak{p}K[X]$ is the prime ideal associated with (x) in $K[X]$. It is clear that $\mathfrak{p}\Omega[X]$ is a prime ideal in $\Omega[X]$, and this shows that the class of regular varieties are precisely those to which nothing bad happens when we change the ground field.

In the rest of this book, the word *variety* will always be understood in the absolute sense, i.e. those coming from a prime ideal $\mathfrak{P}$ in $\Omega[X]$. When we deal with the algebraic set of zeros of a prime ideal $\mathfrak{p}$ in $k[X]$ we shall call this a *k-variety* as in Chapter I.

4. Conjugate varieties

Let K be a field, and $\sigma : K \to K^\sigma$ an isomorphism. If $f \in K[X]$, we can write $f = \sum c_\alpha M_\alpha(X)$ with $c_\alpha \in K$. We denote by f^σ the polynomial $\sum c_\alpha^\sigma M_\alpha(X)$. If $\mathfrak{a}$ is an ideal of $K[X]$ than $\mathfrak{a}^\sigma$ denotes the ideal in $K^\sigma[X]$ consisting of all f^σ as f ranges over $\mathfrak{a}$.

It follows immediately from the definitions that if $\mathfrak{a}$ is the associated ideal of an algebraic set A, then $\mathfrak{a}^\sigma$ is also the associated ideal of an algebraic set which will be denoted by A^σ. Furthermore if $\mathfrak{a} = \mathfrak{p}$ is a prime ideal so is $\mathfrak{p}^\sigma$. If $\mathfrak{p}$ is the associated prime ideal in $K[X]$ of a K-variety V in S^n, then $\mathfrak{p}^\sigma$ is the associated prime ideal in $K^\sigma[X]$ of a K^σ-variety V^σ also in S^n. Let (x) be a generic point of V over K, and let σ be extended in some way to an isomorphism of the field $K(x)$ which we shall also denote by σ. We denote by (x^σ) the point $(x_1^\sigma, \ldots, x_n^\sigma)$. Then (x^σ) is a generic point of V^σ over K^σ and if $K(x)$ is a regular extension of K, then $K^\sigma(x^\sigma)$ is a regular extension of K^σ. Hence if V is a variety, so is V^σ. All these statements are obvious from the definitions.

It is a set-theoretical accident that σ cannot always be extended to an automorphism of the universal domain. If the cardinality of the transcendence degree of Ω over K is the same as that of Ω over K^σ then σ can be so extended. This is easily verified by selecting a transcendence base for Ω over K, extending σ to the field generated over K by this base, and then extending to the algebraic closure. We could counter the set-theoretical difficulty by restricting our family of fields in such a way that σ can always be extended. We leave it to the reader to select his own convention to deal with this set-theoretic oddity, and we assume that σ can be extended to an automorphism of Ω. It always can in practice, when K is finitely generated over a field k left fixed by σ.

Let σ be an automorphism of Ω. Then a point (x) is a zero of an ideal $\mathfrak{a}$ in $K[X]$ if and only if (x^σ) is a zero of the ideal $\mathfrak{a}^\sigma$ in $K^\sigma[X]$. Hence A^σ consists of all points (x^σ), where (x) ranges over A. If we deal with a variety V, and $\mathfrak{P}$ is its associated prime ideal in $\Omega[X]$, then $\mathfrak{P}^\sigma$ is the associated prime ideal of V^σ, and we know that $V^\sigma = V$ if and only if $\mathfrak{P}^\sigma = \mathfrak{P}$. Using Theorem 7 of § 2 we get

THEOREM 9. *Let V be a variety and k_0 its smallest field of definition. Let σ be an automorphism of the universal domain. Then $V^\sigma = V$ if and only if k_0 is left fixed by σ.*

We emphasize the fact that V^σ is entirely determined by the effect of σ on a field of definition of V, and that such a field of definition can be selected finitely generated over the prime field. We do not need an automorphism of the universal domain to be able to define V^σ. However, if we are given such an automorphism σ, and if V is a variety, or A is an algebraic set, then we can define not only V^σ and A^σ but also more precisely the effect of σ on the points of V or of A. The point set of V, or A, is called its support, and we may say that we can define independently V^σ and $(\text{supp } (V))^\sigma$. It turns out of course that $(\text{supp } (V))^\sigma$ is the point set of V^σ and hence we may usually identify V^σ with $(\text{supp } (V))^\sigma$ without fear of confusion.

If V is a variety defined over an algebraic extension K of k, and if σ is an isomorphism of K, then V^σ will be said to be *conjugate* to V over k. The results of Chapter II, § 5 can now be applied in the present circumstances. A k-variety splits up into conjugate varieties over the algebraic closure of k, and hence over a finite algebraic extension of k.

An important special case is that of points. A point (x) is algebraic over k if all its coordinates are algebraic over k. This being so, the conjugates of (x) over k are simply (x^σ) where σ ranges over all isomorphisms of the field $k(x)$ over k. If $[k(x) : k] = n$, then there are at most n points conjugate to (x) over k, because there may be inseparability in the extension. There are exactly n points if and only if the extension $k(x)$ over k is separable. By a *complete set of conjugates* of a point (x) over k, we mean the set of points $((x^{(1)}), \ldots, (x^{(n)}))$ consisting of all conjugates of a given point (x), each one repeated a number of times equal to the inseparable degree of $k(x)$ over k. If x consists of one element, this coincides with the classical concept of conjugates of an element over a field.

THEOREM 10. *Let V be a variety defined over the algebraic closure $\bar{k}$*

of k, and let k_1 be the smallest field of definition of V containing k. Then the number of conjugates of V over k is equal to the degree of separability $[k_1 : k]_s$.

Proof: Let K be a normal extension of k containing k_1, and let G be its group of automorphisms over k. By Theorem 9 an element σ of G leaves V fixed if and only if σ leaves k_1 fixed. The subgroup H of G leaving k_1 fixed has index $[k_1 : k]_s$ in G. This proves our theorem.

5. The Zariski topology

In affine space S^n we define a topology by letting the closed sets be the algebraic sets in S^n. That this is a topology is trivially verified. It is not Hausdorff. Every variety V has an induced topology, and the closed subsets of V are the algebraic sets contained in V. They are the finite unions of subvarieties of V.

The topology has been defined without reference to a field of definition. We shall now prove the equivalence of certain conditions, which will allow us to define a k-topology. We contend that the following conditions concerning an algebraic set A are all equivalent.

C1. A contains all specializations over k of all of its points.

C2. Every component of A is defined over k, and $A^\sigma = A$ for all automorphisms of k over k.

C3. There exists a field of definition K containing k for all the components of A such that $A^\sigma = A$ for all isomorphisms of K over k.

C4. For every automorphism σ of Ω/k, we have $A^\sigma = A$.

C5. For every automorphism σ of Ω/k and every point $(a) \in A$, the point (a^σ) is also in A, or equivalently $(\mathrm{supp}\,(A))^\sigma = \mathrm{supp}\,(A)$.

C6. The smallest field of definition for the ideal $\mathfrak{a}$ associated with A in $\Omega[X]$ is purely inseparable over k.

C7. A is the algebraic set of zeros of equations with coefficients in k (or we can also say of some ideal having a basis in k).

We shall prove all these conditions equivalent by going around in a circle.

C1 $\Rightarrow$ C2. Let $K \supset k$ be a common field of definition for all the components of A, and let (x) be a generic point over K for one of these components, say V. Let V' be the locus of (x) over $\bar{k}$. Then V' is a variety, and by assumption V' is contained in A. Since V is the set of all specializations of (x) over K, which contains k, we see trivially that $V \subset V'$. Since V is a component of A, we must have $V = V'$. This shows that all components of A are defined over $\bar{k}$. Furthermore, if V is a component and σ an automorphism of $\bar{k}$ over k, and if (x) is a generic point of V over $\bar{k}$, then we can extend σ to an isomorphism of $\bar{k}(x)$, and (x^σ) is a generic point of V^σ over $\bar{k}$. Since it is also a specialization of (x) over k (a generic one in fact) it is in A by hypothesis, and hence V^σ, which consists of all its specializations over $\bar{k}$, is also contained in A. This concludes the proof.

C2 $\Rightarrow$ C3. We can take the field of definition algebraic over k and every isomorphism of such a field over k can be extended to an automorphism of $\bar{k}$ over k.

C3 $\Rightarrow$ C4. Every automorphism of Ω over k induces an isomorphism of a field of definition K for all the components of A. So the present statement is almost a restatement of the preceding one.

C4 $\Rightarrow$ C5. We have the general and obvious result that for two algebraic sets $B \subset A$, we also get $B^\sigma \subset A^\sigma$ for any automorphism σ of Ω. Here we take B to be a point (a) in A. It should be remarked that the hypothesis $(\text{supp } (A))^\sigma \subset \text{supp } (A)$ implies $(\text{supp } (A))^\sigma = \text{supp } (A)$, because we can apply σ^{-1}.

C5 $\Rightarrow$ C6. Let K be the smallest field of definition for the ideal $\mathfrak{a}$ associated with A in $\Omega[X]$. Then $(\text{supp } (A))^\sigma$ is the point set of A^σ and its associated ideal is $\mathfrak{a}^\sigma$. By hypothesis they are equal. Hence K must be fixed under all automorphisms of σ over k, and is therefore purely inseparable over k.

C6 $\Rightarrow$ C7. Let $(f_1, \ldots, f_m)$ be a basis for the ideal $\mathfrak{a}$ of A in $\Omega[X]$, with coefficients which lie in a purely inseparable extension of k. Then for a suitable power p^s of the characteristic, $f_j^{p^s}$ has coefficients in k for all j. Clearly, A is the algebraic set of zeros of the equations $f_1^{p^s} = 0, \ldots, f_m^{p^s} = 0$.

C7 $\Rightarrow$ C1. Obvious.

An algebraic set (or a closed set) is said to be *k-closed* if it satisfies any one of the above equivalent properties. It follows immediately that the k-closed sets define a topology on S^n, and that the k-closed subsets of a variety V define a topology on V, which is called the *k-topology*. Every k-closed set is closed, so the k-topology is either finer or coarser than the other (no one is ever able to remember which). The complement of a k-closed set is of course said to be *k-open*.

Note that none of our conditions except the last two refers to any embedding of A in S^n. They will be used later to define a topology on abstract varieties.

A k-closed subset of a variety V is said to be *proper* if it is not equal to V. The complement of a proper k-closed set is a non-empty open set. Note the remarkable fact that two non-empty open sets always have a non-empty intersection!

The maximal chain condition on ideals in a polynomial ring $k[X]$ or $\Omega[X]$ can be translated in terms of the Zariski topology as follows:

PROPOSITION 9. *The k-topology is compact.*

If V is a variety defined over a field k and (x) a generic point of V over k, then $k(x)$ is a regular extension. This has the following consequence, which refines the Hilbert Nullstellensatz.

PROPOSITION 10. *Let V be a variety defined over a field k. Then the set of points of V which are separable algebraic over k (i.e., points (x) such that $k(x)$ is separably algebraic over k) is dense for the k-topology of V.*

Proof: We must show that given a proper k-closed subset A of V, there exists a point of V, separable algebraic over k, and not lying in A. We first make a trivial reduction of the problem. Let (x) be a generic point of V over k, and let $y = f(x)$ be a non-zero element of $k[x]$ which does vanish on A. Such a y exists because $A \neq V$. Consider the point $(x, 1/y)$. A specialization of that point is of type $(x', 1/f(x'))$, and $f(x')$ cannot be 0. Hence (x') cannot be in A. This shows that it suffices to prove that a given variety has points which are separable algebraic over the field k.

Let now (t) be a separating transcendence base of $k(x)$ over k. Then (t) can be viewed as a generic point of affine space S^r if $r = \dim V$. Let y be an element of $k(x)$. We shall prove that the set of points (t') of S^r such that

1. y is integral over the local ring of (t') in $k(t)$, and
2. any specialization y' of y over $(t) \to (t')$ is separable algebraic over $k(t')$,

contains a non-empty k-open set of S^r. In the first place, writing an equation for y with coefficients in $k[t]$, we see that the set of points (t') which do not annul the leading coefficient contains a non-empty k-open set, and for those (t'), y is integral over the local ring. The irreducible polynomial for y over $k(t)$ has no multiple roots because (t) is separating. It can be written $f(t, Y)$. Let $f'(t, Y)$ be its derivative with respect to Y. Then f and f' are relatively prime in $k(t)[Y]$ and after clearing denominators, we can write

$$g_1(t, Y)f(t, Y) + g_2(t, Y)f'(t, Y) = h(t)$$

where $h(t)$ is a polynomial in $k[t]$. Values (t') which do not make $h(t')$ vanish will serve our purpose. Indeed, we know that $f(t, y) = 0$, and if y' is any specialization of y over $(t) \to (t')$ then $f(t', y') = 0$. Furthermore, $f(t', Y)$ has no roots in common with its derivative $f'(t', Y)$. Hence y' is separable over $k(t')$. That (t') can be selected separable over k is clear.

In order to get a point (x') of V which is separable over k, we only need to apply the above arguments to the finite number of coordinates of (x), and use the fact that the intersection of non-empty k-open sets contains a non-empty k-open set. This concludes the proof.

We shall not develop here the general theory of properties which hold on open sets, but we give one more example of a closed set.

PROPOSITION 11. *Let U and V be two varieties defined over a field k, and let F be a k-closed subset of $U \times V$. Then the set of points P on U such that $P \times V \subset F$ is k-closed.*

Proof: We begin by showing that our set A is closed. For each point Q of V we consider the set F_Q of points P in U such that (P, Q) is in F. Then F_Q is closed, being the projection on U of the intersection of F with $U \times Q$. Our set A is the intersection of the F_Q taken over all Q of V, and is therefore closed. To show A is k-closed, we must prove that for every automorphism σ of the universal domain over k, $A^\sigma = A$. But if P is in A, then by hypothesis, $P \times V$ is in F. Applying σ we see that $P^\sigma \times V$ is also contained in F, as desired.

6. Rationality of a cycle over a field

By a *cycle* (of dimension r) on S^n we shall mean an element of the free abelian group generated by the varieties of dimension r. This definition can be generalized to arbitrary varieties, once we have notion of a simple subvariety, which will be defined later. In that case, a *cycle on a variety* V is an element of the free abelian group generated by the simple subvarieties of dimension r on V. The free abelian group generated by all subvarieties of dimension r of V is called the group of r-*chains* on V. For our purposes here, the notion of a cycle on S^n will suffice.

A cycle can be expressed as a formal linear combination $Z = \sum n_i V_i$ of distinct varieties V_i with integer coefficients n_i. By the *support* of the cycle, noted supp (Z), we mean the closed set which is the union of all points which lie in some component V of Z appearing with non-zero coefficient.

We wish to define the notion of rationality of a cycle over a field. Consider an extension $k(x)$, not necessarily regular. For each transcendence base $t_1, \ldots, t_d$ of $k(x)$ over k, the extension $k(x)$ over $k(t)$ is finite algebraic and has a degree of inseparability $[k(x) : k(t)]_i$, which is equal to the degree of $k(x)$ over the maximal separable intermediate field between $k(t)$ and $k(x)$. The minimal value of all these degrees of inseparability as (t) ranges over all possible transcendence bases of $k(x)$ over k is called the *order of inseparability* of $k(x)$ over k. If V is a variety defined over a field k', algebraic over k, and if (x) is a generic point of V over k', then the *order of inseparability of* V *over* k is defined to be that of $k(x)$ over k, which clearly does not depend on the selection of field k'

nor on the generic point (x) of V over k'. The orders of inseparability are denoted by $[k(x) : k]_i$ and $[V : k]_i$. They are equal to a
power of the characteristic.

If σ is an automorphism of the universal domain, and
$Z = \sum n_i V_i$ is a cycle, then we define Z^σ to be the cycle $\sum n_i V_i^\sigma$.

We now define a cycle Z to be *rational over a field k* if the following two conditions are satisfied:

1. For every automorphism σ of the universal domain leaving
k fixed, we must have $Z^\sigma = Z$.

It follows immediately from Condition 1 that $(\text{supp } (Z))^\sigma =$
supp (Z). Hence the support of Z is k-closed, and every component
V appearing in the cycle must be defined over an algebraic extension of k. Our second condition can then be formulated as follows:

2. If V is a component of the cycle, then it appears with
a coefficient divisible by its order of inseparability over k.

The important thing to remember about the power of the
characteristic appearing in the definition of a rational cycle is
that it is the unique power which makes the theorems one wants
to be true actually come out to be true. The proof of this statement
is essentially contained in *Foundations,* and for a further discussion
of this matter, we refer the reader to F - IX$_5$ p. 254.

We shall say that a cycle Z is *prime rational* over a field k if Z
can be written

$$Z = \sum_\sigma p^s V^\sigma,$$

where V is a variety defined over an algebraic extension of k,
the sum is taken over all distinct conjugates V^σ of V over k, and p^s
is the order of inseparability of V over k. The reader will immediately verify that an arbitrary cycle rational over k can be
written uniquely as a linear combination with integer coefficients
of distinct prime rational cycles over k of the same dimension.
If $Z \neq 0$, then $Z = \sum m_j Z_j$, where Z_j is prime rational over k,
$m_j \neq 0$, and all Z_j are distinct.

The support of a prime rational cycle Z over k is precisely a
k-variety in the sense of Chapter I. If we omitted the coefficient
p^s from the expression above, then we would get the finite number

of conjugate varieties into which such a k-variety splits over the algebraic closure of k.

We are interested in the existence of a smallest field of rationality for a cycle, just as we had a smallest field of definition for a variety. It is not always true that such a field exists, as shown by the following example. Let (t, u) be a generic point of the plane S^2 over the prime field k_0 (that is, t and u are two independent variables over k_0). Let $p \neq 0$ be the characteristic, and consider the cycle $p(t, u)$. Then it is rational over $k_0(t, u^p)$ and over $k_0(t^p, u)$ but not over the intersection of these two fields, which is $k_0(t^p, u^p)$ Note however that according to the corollary of Theorem 13 below, there is a smallest field contained in $k_0(t, u^p)$ over which the cycle is rational, namely that field itself.

We shall now give a sufficient condition under which a smallest field of rationality for a cycle exists.

THEOREM 11. *If Z is a cycle on S^n and none of its coefficients are divisible by the characteristic p (if it is > 0) then there exists a smallest field of rationality k_0 for Z. If all components of Z occur with the same multiplicity, this field is the smallest field of definition of the ideal $\mathfrak{a}$ in $\Omega[X]$ belonging to the support of Z, and an automorphism σ of Ω is such that $Z^\sigma = Z$ if and only if σ leaves k_0 fixed.*

Proof: We may write $Z = \sum \mu Z_\mu$ where Z_μ is the sum of the terms with the coefficient μ in the expression of Z as a linear combination of varieties. Then Z is rational over a field if and only if each Z_μ is rational over that field, and an isomorphism which leaves Z invariant must leave the Z_μ invariant. It will therefore be enough to deal with the case where all components of Z appear with the same multiplicity, prime to the characteristic. We may also assume this multiplicity to be positive.

Let k_0 be the smallest field of definition for $\mathfrak{a}$. We shall prove that k_0 is the smallest field of rationality for Z.

Suppose that Z is rational over a field k. We prove that k_0 is contained in k. By definition, Z is fixed under every automorphism of Ω over k. Hence its support is likewise fixed, and this implies that A is k_0-closed. Consider the compositum k_0k. If it is not purely inseparable over k, then there exists an isomorphism of k_0k which is identity on k, but not identity on k_0. This isomorphism cannot map A into itself, and we have a contradiction. Hence k_0k is purely inseparable over k. Furthermore every component of Z is defined over a separable algebraic extension L of k by definition and the hypothesis that the coefficients of Z are not divisible by p. If we can show that $\mathfrak{a}$ is defined over L, then $\mathfrak{a}$ must be defined over $k_0k \cap L = k$, and hence k_0 is contained in k.

To show that $\mathfrak{a}$ is defined over L, we take a polynomial $f(X)$ in $\Omega[X]$ vanishing on A and hence on all components of A. We can write $f = \sum c_j f_j(X)$, where the c_j are elements of Ω linearly independent over L, and $f_j(X) \in L[X]$. If V is a component of A, then V is defined over L by assumption. Let (x) be a generic point of V over L, such that $L(x)$ is linearly disjoint from the field $L(c_j)$ obtained from L by adjoining all coefficients c_j. Then $\sum c_j f_j(x) = 0$. By the linear disjointness, we must have $f_j(x) = 0$. This shows that $f_j(X)$ vanishes on each component of A, and hence on A. From this we conclude that the ideal $\mathfrak{a}$ associated with A in $\Omega[X]$ is defined over L, as desired.

In order to finish the proof, we must show that Z is rational over k_0. Since k_0 is a field of definition for $\mathfrak{a}$ it follows that all components of A (or of Z) are defined over an algebraic extension of k_0. Since A is k_0-closed, all that remains to be shown is that all these components are defined over a separable extension of k_0.

Let $V_1, \ldots, V_m$ be the distinct components of A. Suppose that the smallest field of definition k_1 of V_1 containing k_0 is not separable over k_0. Let k be the maximal separable extension between k_0 and k_1. Then k_1 is purely inseparable over k, and $\mathfrak{a}$ is defined over k. Let $\mathfrak{a}_k$ be the ideal associated with A in $k[X]$, and $\mathfrak{a}_1$ the ideal associated with A in $k_1[X]$. Then $\mathfrak{a}_1 = \mathfrak{a}_k \cdot k_1[X]$. Let $\mathfrak{p}_k$ and $\mathfrak{p}_1$ be the prime ideals in $k[X]$ and $k_1[X]$ respectively vanishing on a generic point (x) of V_1 over k_1. We know by Proposition 8 of § 2 that the ideal $\mathfrak{q} = \mathfrak{p}_k \cdot k_1[X]$ is primary. Since V_1 is a component of A and is defined over k_1, it is *a fortiori* a k_1-component of the

k_1-closed set A, and there is a polynomial g in $k_1[X]$ vanishing on all the other k_1-components of A, but not on V_1 (because there is no inclusion relation among their associated prime ideals in $k_1[X]$).

The assumption that k_1 is the smallest field of definition for V_1 and is not equal to k implies the existence of a polynomial f in $k_1[X]$ which is in $\mathfrak{p}_1$ but not in q. The product fg vanishes on A and hence is in $\mathfrak{a}_1 = \mathfrak{a}_k k_1[X]$. Since $\mathfrak{a}_k \subset \mathfrak{p}_k$ we see that fg is contained in q. But no power of g can be contained in q because g does not vanish on V_1. Since q is primary, we conclude that f must be in q, a contradiction.

The last sentence in our theorem concerning the automorphism is obvious, if we take into account Theorem 7 of § 2. This concludes the proof of the theorem.

If the coefficients of the cycle Z are divisible by the characteristic, then we know that the conclusion of Theorem 11 does not remain valid. However, there is an important case in which it does. In a later chapter we shall define the notion of simple subvariety of codimension 1 on a variety V. The free abelian group generated by these subvarieties is called the group of divisors. It is a theorem of Chow that for divisors the conclusion of Theorem 11 does remain valid without restriction on the coefficients. In fact, we have

THEOREM 12. *If Z is a divisor on a variety V defined over k, then there exists a smallest field of rationality $k_0 \supset k$ for Z, and an automorphism σ of Ω over k leaves Z fixed if and only if σ leaves k_0 fixed.*

This theorem is of considerable importance. The reader will find a proof in Chapter VI, and also another proof in Chow's original paper ("On the defining field of a divisor," Proc. Am. Math. Soc., Vol. 1, No. 6, December 1950, pp. 797—799) where it is shown in addition that this smallest field is generated over the prime field by the Chow coordinates associated with the divisor.

We conclude this section by stating some more useful criteria concerning the rationality of prime rational cycles.

If (x) is a point, then the locus of (x) over k is a k-variety.

The cycle whose components are the components of A, with the multiplicity p^s, where p^s is the order of inseparability of $k(x)$ over k, is called the *prime rational cycle determined by (x) over k.*

THEOREM 13. *Let (x) be a point and let k be a field contained in another field K. Let Z and Z' be the prime rational cycles determined by (x) over k and K respectively. Then $Z = Z'$ if and only if K and $k(x)$ are linearly disjoint over k.*

Proof: In the first part of the proof, we shall quote a result of F - I_8. The reader will see immediately that this result is proved in a very simple manner in *Foundations*, and is self-contained. This is therefore compatible with the point of view taken in this book.

Suppose that $k(x)$ and K are linearly disjoint. Then by Theorem 8 of § 2 the ideal determined by (x) in $K[X]$ has a basis in $k[X]$. Hence supp (Z) = supp (Z'). This implies that the components of Z and Z' are the same. Again using the fact that $k(x)$ and K are linearly disjoint over k, one sees that the order of inseparability of (x) over k is the same as that of (x) over K (F - I_8 Prop. 26). Hence the components of Z and Z' occur with the same multiplicity in both cycles, as was to be shown.

Conversely, suppose $Z = Z'$. Then supp (Z) = supp (Z') and we see that K and $k(x)$ must be free over k. Let k_1 be the separable algebraic closure of k in $k(x)$. Then the number of distinct components in Z' is $[k_1 : k]_s$ by Theorem 10 of § 3. This must also be the number of distinct components in Z, and hence $[k_1 : k] = [Kk_1 : K]$. Hence k_1 and K are linearly disjoint over k. Let (u) be a transcendence base of $k(x)$ over k such that the degree of inseparability $[k(x) : k(u)]_i$ is equal to the order of inseparability of $k(x)$ over k. Then $k(u)$ and K are linearly disjoint over k by Proposition 3 of § 1. By assumption, the order of inseparability of $k(x)$ over k is the same as that of $K(x)$ over K. If we put $K_1 = Kk_1$, then $K_1(u)$ and $k_1(x)$ must be linearly disjoint over $k_1(u)$, and by Proposition 1 of § 1 we conclude that $k(x)$ and K are linearly disjoint. This proves our theorem.

COROLLARY. *Let Z be a prime rational cycle over a field K*

and let k be contained in K. If Z is rational over k, then Z is prime rational over k, and there exists a smallest field k_0 contained in K over which Z is rational (hence prime rational). If Z is determined by the point (x) over K, and if $\mathfrak{P}$ is the prime ideal of $K[X]$ vanishing on (x), then this smallest field k_0 is the smallest field of definition of $\mathfrak{P}$. If σ is an automorphism of K, then σ leaves Z fixed if and only if σ leaves k_0 fixed.

Proof: If V and V' are two components of Z, then they are conjugate over K, and *a fortiori* over k. Since Z is rational over k, the set of all conjugates of V over k coincides with the set of all conjugates of V over K. All that remains to be proved is that the order of inseparability of V over k is the same as that of V over K. Let (x) be a generic point of V over k. Then it is immediate from the definition of order of inseparability that $[k(x) : k]_\iota \geqq [K(x) : K]_\iota$. Under the hypothesis that Z is prime rational over k, these must be equal. The rest of the theorem is now obvious, if we take into account Theorem 7 of § 2, Theorem 8 of § 2, and the theorem that has just been proved.

Literature

In the first section of this chapter we attempt to make a complete list of the basic theorems concerning separable, regular, and primary extensions. These are used constantly in the theory of algebraic systems, especially by Chow in his theory of algebraic systems of abelian varieties [1].

The existence of a smallest field of rationality for a cycle (or a divisor) is extremely important. In many questions, it replaces the use of the Chow coordinates (compare for instance Matsusaka's methods [2, 3] with Weil's [4, 5]).

CHOW, W. L.

[1] Abelian varieties over function fields, Trans. Am. Math. Soc. Vol. 78, 1955, pp. 253—275.

MATSUSAKA, T.

[2] On the algebraic construction of the Picard Variety, Japanese journal of Math. Vol. XXI, 1951, pp. 217—235.

[3] Some theorems on abelian varieties, Natural Science Report, Ochanomizu University, Vol. 4, No. 1, 1953, pp. 22—34.

WEIL, A.

[4] Algebraic groups of transformations, Am. J. of Math., Vol. 77, No. 2, April 1955, pp. 355—391.

[5] The smallest field of definition of a variety, Am. J. of Math., Vol. 78, No. 3, July 1956, pp. 509—524.

CHAPTER IV

Products, Projections, and Correspondences

This chapter consists mostly of definitions, and we do not prove anything essentially new. The dimension theorem of Chapter II has an interesting corollary concerning the blowing up of varieties under correspondences (Theorem 2 below).

If the reader assumes that the ground field is fixed and algebraically closed, then this chapter does not depend on Chapter III (provided one disregards all references to changes of fields of definition).

One more warning: The word "projection" is used in a technical sense in algebraic geometry, and does not coincide with the set theoretic projection. The latter will be seen to contain an open (non-empty) subset of the former.

1. Products of varieties

We consider affine space S^n with variables $(X) = (X_1, \ldots, X_n)$ and affine space S^m with variables $(Y_1, \ldots, Y_m)$. Let V be a variety in S^n, defined over field k, and let $\mathfrak{p}$ be its prime ideal in $k[X]$. Let W be a variety in S^m also defined over k, and with prime ideal $\mathfrak{q}$ in $k[Y]$. Let (x) and (y) be generic points of V and W over k. We can always find two such points such that (x) and (y) are free over k (i.e. $k(x)$ and $k(y)$ are free over k). Since the extension $k(x)$ is regular over k, and $k(y)$, it follows that $k(x)$ and $k(y)$ are linearly disjoint over k by Theorem 3 of Chapter III, § 1. Furthermore, $k(x, y)$ is a regular extension of $k(y)$ by Corollary 4 of Theorem 2, Chapter III, § 1, and hence $k(x, y)$ is a regular extension of k. The locus of (x, y) over k is therefore a variety.

On the other hand, we define $V \times W$ to be the set of all pairs (x', y') with (x') in V and (y') in W. It is clear that $V \times W$ is the algebraic set (k-closed) of zeros of the ideal $(\mathfrak{p}, \mathfrak{q})$ in $k[X, Y]$. The following proposition shows that $V \times W$ is a variety that

($\mathfrak{p}$, $\mathfrak{q}$) is the prime ideal associated with $V \times W$ in $k[X, Y]$, and that (x, y) is a generic point of $V \times W$ over k.

PROPOSITION 1. *Let (x) and (y) be two points in S^m and S^m respectively such that the fields $k(x)$ and $k(y)$ are linearly disjoint. Let $\mathfrak{p}$ be the ideal of (x) in $k[X]$ and $\mathfrak{q}$ the ideal of (y) in $k[Y]$. Then the ideal $(\mathfrak{p}, \mathfrak{q})$ in $k[X, Y]$ is the prime ideal of $k[X, Y]$ vanishing on (x, y).*

Proof: We must prove that $(\mathfrak{p}, \mathfrak{q})$ is the kernel of the map $f(X, Y) \to f(x, y)$. It is obviously contained in the kernel. Suppose $f(x, y) = 0$. We regard $k[X]$ as a vector space over k, and we let $\{M_\beta(X)\}$ be a maximal set of monomials in (X) linearly independent mod $\mathfrak{p}$. Then the $\{M_\beta(x)\}$ form a basis for $k[x]$ over k. Similarly, we let $\{N_\gamma(Y)\}$ be a maximal set of monomials in (Y) linearly independent mod $\mathfrak{q}$. Then the $\{N_\gamma(y)\}$ form a basis of $k[y]$ over k. Write $f(X, Y)$ as a linear combination of monomials in (X) with coefficients in $k[Y]$. Let $R = k[X, Y]$. Then

$$f \equiv \sum g_\beta(Y)M_\beta(X) \qquad (\mathrm{mod}\ \mathfrak{p}R).$$

Write each $g_\beta(Y) \equiv \sum_\gamma a_{\beta, \gamma} N_\gamma(Y)$ (mod $\mathfrak{q}$), with $a_{\beta, \gamma} \in k$. Then

$$f(X, Y) \equiv \sum_\beta (\sum a_{\beta, \gamma} N_\gamma)M_\beta \qquad (\mathrm{mod}\ (\mathfrak{p}, \mathfrak{q})R).$$

We contend that all $a_{\beta, \gamma} = 0$. Substitute (x, y) for (X, Y). We get

$$0 = f(x, y) = \sum_\beta \sum_\gamma (a_{\beta, \gamma} N_\gamma(y))M_\beta(x).$$

The linear disjointness of $k(x)$ and $k(y)$ shows that all coefficients $a_{\beta, \gamma} = 0$, thereby proving our theorem.

COROLLARY 1. *Let K and L be two extensions of k which are linearly disjoint over k. Let R be a subring of K, and R' a subring of L. Let φ be a homomorphism of R into an algebraically closed field F, and let φ' be a homomorphism of R' into F. Then there exists an F-valued place ψ of KL whose restriction to R is φ and to R' is φ'.*

Proof: If R and R' are finitely generated over k, then we can write $R = k[x]$ and $R' = k[y]$ and by the theorem, the two homomorphisms of R and R' induce a homomorphism of $k[x, y]$ into F, which can be extended to an F-valued place. The infinite case is now an immediate consequence of the finite case.

COROLLARY 2. *If V is a variety of dimension r, and W a variety of dimension s, then $V \times W$ has dimension $r + s$.*

Proof: If (x, y) is a generic point for $V \times W$, then its transcendence degree over k is $r + s$.

PROPOSITION 2. *Let V and W, be as above, two varieties. Let k and k' be the smallest field of definition for V and W respectively. Then the compositum kk' is the smallest field of definition for $V \times W$.*

Proof: It is clear that kk' is a field of definition for $V \times W$. Conversely, let K be such a field of definition, and let L be the field Kkk' obtained by composing the three fields. Then L is a field of definition for V and W, and all we need to show is that $L \subset K$. Let (x, y) be a generic point of $V \times W$ over L. It is also a generic point of $V \times W$ over K, and we know by Theorem 8 of Chapter III, § 2 that $K(x, y)$ and L are linearly disjoint over K, and hence free. The point (x) is a point of V, and since $K(x)$ is contained in $K(x, y)$, it is a regular extension of K. Hence the locus of (x) over K is a variety, which is contained in V because the locus of (x) over K is the same as over L which is a field of definition for V. Furthermore, $\dim_K(x) = \dim_L(x) \leqq \dim V$. We get similar results for (y) and W. We can therefore conclude that these dimensions must be equal, because $\dim_K(x, y)$ must be $r + s$. This proves that the locus of (x) over K is V, and that K is a field of definition for V. Similarly for W. This concludes the proof.

2. Projections

PROPOSITION 3. *Let (x, y) be a point in $S^n \times S^m$, and let $\mathfrak{P}$ be the prime ideal in $k[X, Y]$ vanishing on that point. Then the ideal $\mathfrak{p} = \mathfrak{P} \cap k[X]$ is the prime ideal in $k[X]$ vanishing on (x), and (x) is a generic zero of $\mathfrak{p}$.*

Proof: It is obvious that (x) is a zero of $\mathfrak{p}$. Conversely, let (x') be a generic zero of $\mathfrak{p}$. We shall prove that $(x) \to (x')$ is a specialization, thereby proving that (x) is also generic. Let $f(X) \in k[X]$ and suppose $f(x) = 0$. Then f vanishes on (x, y) *a fortiori*, and hence $f \in \mathfrak{P}$, so $f \in \mathfrak{p}$. Hence $f(x') = 0$. This proves what we wanted.

Let now U be a variety in a product space $S^n \times S^m$, let k be a field of definition for U, and let (x, y) be a generic point of U over k. The variety U' in S^n determined by the ideal $\mathfrak{p}$ in $k[X]$ and having the generic point (x) over k will be called the *projection of U on the first factor*, or on S^n. If U is contained in a variety $V \times W$ in $S^n \times S^m$, it will also be called the *projection of U on V*. If we work over the universal domain Ω, then the ideal vanishing on U' in $\Omega[X]$ is the intersection with $\Omega[X]$ of the ideal vanishing on U in $\Omega[X, Y]$. It is clear that the projection of U on V is contained in V. If k is a field of definition for U, then it is also a field of definition for its projection on the first factor. If the projection of U is a subvariety A of V, we shall also say that U *lies above* A.

In general the projection of U in the above sense does not coincide with the set theoretic projection on the first factor. As an example, we let U be the variety defined by the equation $X_1^2 + X_2 Y = 0$. The projection of U on S^2 is the entire space, but the set theoretic projection consists of $(0, 0)$ and of all points (x_1, x_2) such that neither x_1 nor $x_2 = 0$, or such that $x_1 = 0$ and x_2 is arbitrary. This is not the whole space. However, we still have the following result.

PROPOSITION 4. *Let U be a subvariety of a product $S^n \times S^m$ and let V be its projection on the first factor. Let k be a field of definition for U. Then the set theoretic projection of U on V contains a non-empty open subset of V, and the union of all such sets is k-open. Hence the set theoretic projection of U is k-dense in the projection.*

Proof: Let (x, y) be a generic point of U over k. We must show that there is an open set of points (x') on the locus V of (x) over k for which the specialization $(x) \to (x')$ can be extended to a specialization $(y) \to (y')$, because in that case (x', y') lies in W, and (x') is in the set theoretic projection. By the Hilbert Nullstellensatz we can find a specialization (w) of (y) over $k(x)$ which is algebraic over $k(x)$. For each w_j we can write an algebraic equation with coefficients in $k[x]$. Let $G(x)$ be the product of the leading coefficients. The set of points (x') on V such that $G(x') \neq 0$ is a k-open subset of V. For any such point, we can extend the specialization $(x) \to (x')$ over k to a place φ of $k(x, w)$ which is necessarily finite on (w) because the w_j are integral over the local ring of (x') in $k(x)$. We put $\varphi(w) = (y')$, and it is then clear that the point (x', y') satisfies our requirements. The union of all open sets of V contained in the set theoretic projection of U on V is invariant under all automorphisms of the universal domain leaving k fixed, and is consequently k-open.

3. Rational maps

Let T be a subvariety of a product $V \times W$ with the projection V on the first factor. This means that if (x, y) is a generic point of T over some field of definition k for T, V and W then (x) is a generic point of V, and (y) is a point of W. Suppose that $k(y)$ is contained in $k(x)$. Then it is easily seen by using the existence of a smallest field of definition for T, and the properties of linear disjointness that this property (namely $k(y)$ contained in $k(x)$) is independent of the field of definition k chosen, and of the generic point (x, y) for T over k. We shall then call T the *graph of a rational map $f : V \to W$ of V into W*. We often identify the rational map with its graph and we often call T itself the rational map. We also write $T : V \to W$. If at the same time the projection of T on the second factor is W, then we say that the rational map is *generically surjective*. This means that (y) is a generic point for W over k. We say that k is a field of *definition for f* if it is a field of definition for T. It is then a field of definition for the projections of f on both the first and second factor.

The reader should not view a rational map as giving a mapping of the points of V into those of W, but rather as a mapping of the algebraic subsets of V into those of W according to the following definition. If A is a closed subset of V then we denote by $T(A)$, or by $f(A)$, the projection on W of $T \cap (A \times W)$. Similarly, if B

is a closed subset of W, we denote by $f^{-1}(B)$, or $T^{-1}(B)$, the projection on V of $T \cap (V \times B)$. We call $T(A)$ the *image of A under T*, and $T^{-1}(B)$ the *inverse image* of B.

If we are given a finitely generated extension $k(x)$ of k (with $k(x)$ regular over k, so (x) can be viewed as the generic point of a variety V), then a rational map determines a model of a subfield of $k(x)$, namely $k(y)$. Conversely, given a subfield of $k(x)$, it is finitely generated, by Proposition 6 of Chapter III, § 2 and if we write it as $k(y)$, then (x, y) is the generic point of a variety T, which gives a rational map. Thus a rational map is completely determined by taking a finitely generated regular extension K of a field k, a model V of K, and model W of a subfield E of K, together with generic points (x) and (y) of V and W such that $K = k(x)$ and $E = k(y)$. We have $f(x) = (y)$, or $T(x) = (y)$ according to the preceding definition.

It is easily seen, using the properties of linear disjointness, and Corollary 6 of Theorems 1, 2, and 5 of Chapter III, § 1 that the following properties of a rational map do not depend on the selection of field of definition nor of generic point:

$k(x)$ is a finite extension of $k(y)$, of degree n;

$k(x)$ is a separable extension of $k(y)$;

$k(x)$ is a purely inseparable extension of $k(y)$;

$k(x)$ is a primary extension of $k(y)$;

$k(x)$ is a regular extension of $k(y)$.

If the first condition is satisfied, then we say that the rational map f is of *finite degree*, and that its degree is n. In the other cases, we say, respectively, that f is *separable, purely inseparable, primary*, or *regular*.

PROPOSITION 5. *Let $f : V \to W$ be a rational map, which is generically surjective, and let k be a field of definition for f. Let (x, y) be a generic point for the graph of f. Then $f^{-1}(y)$ is the locus of (x) over $k(y)$.*

Proof: Obvious from the definition.

COROLLARY. *$f^{-1}(y)$ consists of only a finite number of points, if and only if f is of finite degree; it consists of one point only if* *and only if f is purely inseparable, and it is a variety if and only if f is primary.*

If we had intersection theory, then we could define $f^{-1}(y)$ as a cycle. In that case, one can state that f is regular if and only if the cycle $f^{-1}(y)$ is a variety. We know of course that if f is primary, then the support of $f^{-1}(y)$ is a variety, but if f is not regular, then the unique component entering in this cycle will appear with a multiplicity equal to the order of inseparability of $k(x)$ over $k(y)$.

It can be shown fairly easily that if any one of the above properties holds then it holds almost everywhere, i.e. for at least all points (y') on a non-empty k-open subset of W. Some of these properties hold exactly on a k-open set. Unfortunately there is at the time this book is being written no systematic exposition of the type of problem just mentioned. To provide such an exposition would take us considerably beyond the scope of this book, since it is best handled simultaneously with intersection theory, and with the theory of the Chow coordinates.

If a rational map is of finite degree, and separable, then its degree is the number of points lying in the inverse image of a generic point. In view of the above remark, it would also be the number of points lying in the inverse image of almost all points. If the rational map is of finite degree but not separable, then we have taken a definition which is such that we could still say that the degree is the number of points in the inverse image of a generic point, provided that they are counted with suitable multiplicity, namely the degree of inseparability of $k(x)$ over $k(y)$. From Proposition 5, we see that in any case, the points in $f^{-1}(y)$ are all conjugate over $k(y)$.

There is an important special case of rational maps. We shall say that a subvariety T of a product $V \times W$ is a *birational* map (or *birational correspondence*) between V and W if its projection on V is V, its projection on W is W, and if it is a rational map in both directions of degree 1. In terms of generic points, this means that if (x, y) is a generic point of T over a field of definition k for T, then (x) is a generic point of V, (y) is a generic point for W over k, and $k(x, y) = k(x) = k(y)$. We have $T(x) = (y)$, and $T^{-1}(y) = (x)$. To say that we have a birational map amounts therefore to taking two sets of generators of a field $k(x)$.

Let us return to a rational map T of V generically onto W. Let (x') be a point of V, and consider the following statement.

If k is a field of definition for T, and (x, y) is a generic point of T over k, then all coordinates y_j of (y) lie in the local ring of (x') in $k(x)$.

We shall prove that if the above statement is true for one field of definition of V, and some generic point over that field, then it is true for every field of definition. This being the case, if the situation described in the above statement occurs for a point (x') of V, then we shall say that the rational map is *defined* at (x') or *holomorphic* at (x'). We can then write $y_j = f_j(x)/g_j(x)$, $g_j(x') \neq 0$, and (y) has a uniquely determined specialization over $(x) \to (x')$, namely $(y') = f_j(x')/g_j(x')$. In particular, $T \cap ((x') \times W) = (x', y')$, and $T(x') = (y')$ consists of a single point.

Let us now prove the independence of our property on the selection of field of definition and generic point. Once k has been selected, then two generic points differ by an isomorphism, and hence our statement is obviously independent of the choice of (x, y) over k. Denote the statement by $D(k)$, and let k_0 be the smallest field of definition for T. It will suffice to prove that $D(k)$ is true if and only if $D(k_0)$ is true.

Suppose $D(k_0)$ is true. Let (x, y) be a generic point for T over k, and hence over k_0. Then $k_0(x, y)$ is linearly disjoint from k over k_0. An expression of y_j in the local ring of (x') in $k_0(x)$ gives *a fortiori* an expression of y_j in the local ring of (x') in $k(x)$. Hence going from k_0 to k is trivial.

To go in the other direction, we have the following proposition.

PROPOSITION 6.   *Let (x) be a point, and $\mathfrak{o}$ the local ring of a point (x') in the field $k(x)$. Let K be an extension of k linearly disjoint from $k(x)$ over k, and let $\mathfrak{o}_K$ be the local ring of (x') in $K(x)$. Then $\mathfrak{o} = \mathfrak{o}_K \cap k(x)$.*

Proof: The statement of our proposition depends of course

on the fact that (x') is also a specialization of (x) over K, according to Theorem 8 of Chapter III, § 2. Now let y be an element of $k(x)$ which is in $\mathfrak{o}_K$, so we can write $y = f(x)/g(x)$, $f(X)$, $g(X) \in K[X]$ and $g(x') \neq 0$. Let $\{w_\alpha\}$ be a linear base of K over k, and write

$$f(X) = \sum w_\alpha f_\alpha(X), \qquad g(X) = \sum w_\alpha g_\alpha(X)$$

where f_α and g_α have coefficients in k. We get

$$\sum w_\alpha (g_\alpha(x)y - f_\alpha(x)) = 0.$$

Since K and $k(x)$ are linearly disjoint, we must have for every index α, $g_\alpha(x)y - f_\alpha(x) = 0$. Since $g(x') \neq 0$, there is some index α such that $g_\alpha(x') \neq 0$. For that α, we have *a fortiori* $g_\alpha(x) \neq 0$. We have also $y = f_\alpha(x)/g_\alpha(x)$, and this proves our proposition.

Proposition 6 clearly shows that $D(k)$ implies $D(k_0)$. Note that if $\mathfrak{m}$ is the maximal ideal of $\mathfrak{o}$, and $\mathfrak{m}_K$ is the maximal ideal of $\mathfrak{o}_K$ in the proposition, then one sees immediately that $\mathfrak{m} = \mathfrak{m}_K \cap k(x)$. The statements dealing with the extension of $\mathfrak{m}$ to $\mathfrak{o}_K$ will be dealt with in Chapter VI.

Let $T : V \to W$ be a rational map. Let U be a subvariety of V. We shall say that T is *defined at* U if there is some point of U at which T is defined. If that is the case, then one sees that whenever k is a field of definition for T, and U, and (x') is a generic point of U over k, then T is defined at (x'). This comes from the obvious fact that if T is defined at a point (x''), and if (x'') is a specialization of (x') over k, then T is defined at (x'). (Cf. Chapter I, § 3, where we have already met a special case of the present notion.)

Let $T : V \to W$ be a rational map of V into W, defined over k. Let (x, y) be a generic point of T over k, and let (x') be a point of V at which T is defined. Put $T(x') = (y')$. Then $k(y') \subset k(x')$. If the extension $k(x')$ is regular, then (x') can be viewed as the generic point of a subvariety V' of V, and (x', y') is then the generic point of a subvariety T' of $V' \times W'$ where W' is the locus of (y') over k. T' is obviously a rational map of V' into W' (in fact generically surjective), which is called the map *induced by* T *on* V'

We say that T is *everywhere defined* if it is defined at every point. An everywhere defined rational map can be viewed as an ordinary function of V into W.

PROPOSITION 8. *Let $T : V \to W$ be a rational map, defined over a field k. Then the set of points where T is not defined is a proper k-closed subset of V.*

Proof: Let (x, y) be a generic point of T over k, so that by hypothesis, (x) is a generic point of V over k, and $(y) = T(x)$. Let $(y) = (y_1, \ldots, y_m)$. We may view each $(x, y_j)(j = 1, \ldots, m)$ as the graph of a rational map of V into the affine line S^1. The set of points where T is not defined is then the finite union of the sets of points where y_i is not defined. Hence it suffices to prove that the latter are k-closed, i.e. we may restrict our attention to the case where (y) consists of one element of the field $k(x)$. We represent y in all possible ways as a quotient of polynomials $f_\alpha(x)/g_\alpha(x)$ for a suitable indexing set $\{\alpha\}$. Let $\mathfrak{a}$ be the ideal generated in $k[X]$ by all polynomials $\{g_\alpha(X)\}$ appearing in these representations. Then the set of points where y is not defined is the algebraic set of zeros of $\mathfrak{a}$ intersected with V, and this is k-closed.

If T is a birational map between V and W, if (x') is a point of V at which T is defined, put $T(x') = (y')$. If T^{-1} is defined at (y'), then one sees from the definitions that $T^{-1}(y') = (x')$. We shall say in this case that T is *biholomorphic* at (x') and (y') (or simply at (x'), since (y') is then uniquely determined). As a corollary to Proposition 8 we get

COROLLARY. *Let T be a birational map between two varieties V and W, all these objects being defined over k. Then the set of points on V where T is not biholomorphic is k-closed.*

Proof: We leave the proof to the reader as an exercise.

REMARK ON NOTATION. In some books (for instance in *Foundations*) a rational map is said to be "regular" at a point if it is defined at that point in our terminology. This conflicts with the use of the word regular with respect to regular extensions. This is the reason why we have changed the terminology.

If T is a birational map between V and W and is biholomorphic at every point of V and W then we shall say that T is an *isomorphism*. If k is a field of definition for T, V, and W, we say that T is a *k-isomorphism*.

PROPOSITION 9. *Let $T : V \to W$ be an everywhere defined rational map, and let k be a field of definition for T, V, and W. Then T is continuous for the k-topology.*

Proof: Let A be a k-closed subset of W. Then its inverse image is the set theoretic projection on V of $T \cap (V \times A)$, and we must prove that this set theoretic projection coincides with the projection. We consider here the k-components of the intersection as in Chapter II. Let (x, y) be a generic point for one of these k-components. If (x') is a specialization of (x) over k, then $T(x') = (y')$ is defined by hypothesis, and is in A because A is k-closed. Hence (x') is in the projection of this k-component. This shows that the set theoretic projection coincides with the other.

In case T is not everywhere defined, one cannot speak of continuity. On the other hand, we know that T is defined on a non-empty open subset of V, and that if T is generically surjective, then the image of T contains a non-empty open subset of W by Proposition 4 of § 2.

PROPOSITION 10. *Let $f : U \to V$ and $g : V \to W$ be two rational maps. Suppose that f is defined at a point (u') of U and put $f(u') = (v')$. Suppose that g is defined at (v') and put $g(v') = (w')$. Let k be a field of definition for f, U, V, and W. Let (u) be a generic point over k, and let $(v) = f(u)$. Then g is defined at (v). If we put $g(v) = (w)$, then there is a rational map $h : U \to W$ of U into W, such that $h(u) = (w)$. This map h is defined at (u'), and $h(u') = (w')$*

Proof: This is an immediate consequence of the definitions. The rational map h of the proposition is denoted by $g \circ f$.

PROPOSITION 11. *Let $f : U \to V$ and $g : V \to W$ be two generically surjective rational maps. Let k be a common field of definition for f, g, U, V, and W. Let (u) be a generic point of U over k. Put $f(u) = (v)$ and $g(v) = (w)$. If W is equal to U, and $g \circ f$ is identity, then f and g are birational. If furthermore f and g are everywhere defined then they are biholomorphic.*

Proof: By definition, we must have $k(u) \supset k(v) \supset k(w)$. If $g \circ f$ is identity, all three fields are equal, so f and g are birational. One sees immediately from the definitions that f and g must then be biholomorphic.

We conclude this section with one more lemma concerning k-closed and k-open sets, which is a little less trivial than the others, and is used by Weil in his treatment of groups of transformations.

PROPOSITION 12. *Let* U, V, W *be three varieties, and* $f : U \times V \to W$ *a rational map of* $U \times V$ *into* W, *all defined over* k. *Assume that for every* $(a) \in U$, f *is defined at* (a, x) *for* (x) *generic on* V *over* $k(a)$. *Let* B *be the set of those* (a) *in* U *such that, for* (x) *generic over* $k(a)$, $f(a, x)$ *is generic over* $k(a)$ *on* W. *Then* B *is* k-open.

Proof: We begin by proving that if a generic point (u) of U over k is not in B, then B is empty. Let (a) be a point in U. Let (x) be a generic point of V over $k(a)$. We must show that $f(a, x)$ cannot be generic on W over $k(a)$. Let (u) be a generic point of U over $k(a, x)$. By hypothesis, (u) is not in B. Let $(w) = f(a, x)$. Then (w) is a generic point of W over k, but cannot be generic for W over $k(u)$. Say $w_1, \ldots, w_s$ are coordinates of (w) which are independent over k but algebraically dependent over $k(u)$. An algebraic relation between them can be written

$$\sum z_j M_j(w) = 0, \qquad\qquad z_j \in k[u],$$

where the M_j are monomials. Take a place of $k(u)$ which extends the specialization $(u) \to (a)$, and is $\overline{k(a)}$-valued. Then for one of the z_j, say z_1, the quantities $t_j = z_j/z_1$ are all finite under the place, which induces therefore a specialization $(u, t) \to (a, t')$ where the coordinates of (t') are in $\overline{k(a)}$. We extend our place to a place φ of $k(x, u)$, leaving (x) fixed. This can be done because $k(x)$ is linearly disjoint from $k(u)$ over k (cf. Corollary of Proposition 1, § 1). We divide our relation by z_1, to get $\sum t_j M_j(w) = 0$, and apply the place. This yields $\sum t'_j M_j(w) = 0$, because $f(a, x)$ is defined and equal to (w) by hypothesis. This contradicts the fact that the w's are algebraically independent over $k(a)$.

We may from now on assume that a generic point of U over k

lies in our set B. Our next task is to prove that B then contains an open set.

Let dim $V = r$, and let V be in S^n. Let $w_1, \ldots, w_s$ be a transcendence base of $k(u, w)$ over $k(u)$, and let $x_1, \ldots, x_{r-s}$ be a transcendence base of $k(u, x)$ over $k(u, w)$. Denote the elements $(w_1, \ldots, w_s, x_1, \ldots, x_{r-s})$ by $(t) = (t_1, \ldots, t_r)$.

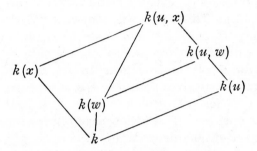

Each t_j can be written as a rational function

$$t_j = F_j(u, x)/G_j(u, x),$$

where F_j, G_j are polynomials in $k[u,x]$, and such that for any point (a, x^*) of $U \times V$ with (x^*) generic over $k(a)$, $G_j(a, x^*)$ is defined. For the $w_1, \ldots, w_s$ this is the hypothesis and for the x_j we take $F_j(u, x) = x_j$ itself and $G_j = 1$.

Each coordinate x_i of (x) $(i = 1, \ldots, n)$ satisfies an algebraic equation with coefficients in $k[u, t]$. In order to have only one equation to write down, we take the product of all of them. Then each x_i satisfies

$$H_d(u, t)x_i^d + \ldots + H_0(u, t) = 0$$

with $H_\nu(u, t) \in k[u, t]$ and $H_d(u, t) \neq 0$.

If we substitute (a, x^*) in this equation, then all denominators $G_j(a, x^*)$ are defined and since the H_ν are polynomials in $k[u, t]$ they are also defined. We contend that there is a non-empty k-open subset Γ of U, such that for $(a) \in \Gamma$ and (x^*) generic over $k(a)$, the expression

$$\left(a, \frac{F_1(a, x^*)}{G_1(a, x^*)}, \ldots, \frac{F_r(a, x^*)}{G_r(a, x^*)} \right)$$

does not vanish. Suppose this contention is proved. Then we see that $k(a, x^*)$ is algebraic over $k(a, t^*)$, where (t^*) is the well-defined point obtained by substituting (a, x^*) for (u, x) in the expression for (t). This implies that $(w_1^*, \ldots, w_s^*)$ must have dimension s over $k(a)$. Since $f(a, x^*)$ is defined by hypothesis, and since its first coordinates are $w_1^*, \ldots, w_s^*$, it follows that $f(a, x^*)$ has dimension s over $k(a)$, and is therefore a generic point of W over $k(a)$. This will finish the second part of our proof.

Multiplying $H_a(u, t)$ by the product $(G_1(u, x) \ldots G_r(u, x))^m$ where m is sufficiently large, we see that it suffices to prove our contention for a polynomial $H(u, x)$ in $k[u, x]$. In other words, we must prove that given $H(u, x) \in k[u, x]$, the set of points (a) of U such that for (x^*) generic over $k(a)$, $H(a, x^*)$ is non-zero contains a k-open set. This is now easy to prove from considerations of linear disjointness. Write $H(u, x) = \sum_{\beta=1}^m H_\beta(u) M_\beta(x)$, where $M_\beta(x)$ are monomials in (X), the $M_\beta(x)$ are linearly independent over k, and $H_\beta(u)$ are non-zero elements of $k[u]$. If (x^*) is generic over $k(a)$, then $k(x^*)$ and $k(a)$ are linearly disjoint over k by Theorem 8 of Chapter III, § 2, and hence $\{M_\beta(x^*)\}$ are linearly independent over $k(a)$. Consequently $H(a, x^*) = 0$ if and only if $H_\beta(a) = 0$ for all β. The set of (a) in U such that $H_\beta(a) = 0$ for all β is k-closed and proper. This ends the second part of our proof.

The third part of our proof consists in proving that B is k-open, and we shall prove our proposition by induction on the dimension of U, exactly as in Weil. Assume that B is not empty, and put $D = U - B$. We have proved that D is contained in a k-closed subset C of U. Call U_i the components of C. They are defined over $\bar{k}$, and of dimension $< \dim U$. By the induction assumption, $B \cap U_i$ is a $\bar{k}$-open subset of U_i, and its complement C_i on U_i is a $\bar{k}$-closed subset of U. As D is the union of the C_i, this shows that D is $\bar{k}$-closed. As it is obviously invariant by all automorphisms of $\bar{k}$ over k, it must be k-closed, and this proves our proposition.

4. Functions and function fields

Let T be the graph of a rational map f of V into S^1, the affine space of dimension 1 (also known as the affine line). Let k be a field of definition for T, and let (x, z) be a generic point of T over k. Then by definition, $z = f(x)$ is an element of $k(x) = K$.

Conversely, let (x) be a generic point of a variety V over a field k, and let z be an element of $k(x)$. Then (x, z) is the generic point of the graph of a rational map of V into S^1.

If we work with a fixed ground field k, then all fields $k(x)$ generated by a generic point of V are isomorphic, in a natural way. The elements of $k(x)$ are called functions of V in $k(x)$. In general, a rational map f of V into S^1 will be called a *function on V*. Defining functions in terms of a subvariety of a product has the advantage that it shows that the concept is geometric, i.e. independent of the ground field. It provides an equivalence relation for elements of fields of type $k(x)$ where (x) is a generic point of V over k, namely, given $z \in k(x)$, and given another field of definition k' for V, a generic point (x') for V over k' and an element z' in $k'(x')$, we can say that z and z' are equivalent if $f(x) = z$ and $f(x') = z'$.

The set of functions on V is a field under the following operations. Let f, g be two functions on V, i.e. rational maps into S^1. Let k be a common field of definition for them. Let (x) be a generic point of V over k, and let $f(x) = z$, and $g(x) = w$. Then $t = z + w$ is an element of $k(x)$, and (x, t) is the graph of a rational map $h : V \to S^1$ of V into S^1. It is easily seen that h does not depend on the choice of field of definition k or of generic point (x) selected. We then define $f + g$ to be h. Similarly, we define $f - g$ and fg, and f^{-1}. It is trivially verified that the set of all functions on V form a field, which is called the *function field of V*, and is denoted by $\Omega(V)$. If k is a field of definition for V, then the subfield of $\Omega(V)$ consisting of those functions defined over k is denoted by $k(V)$ and is canonically isomorphic to $k(x)$ over k, under the mapping $f \to f(x)$.

We sometimes call $k(x)$ a function field of V over k, and its elements functions. We do this mostly when we deal with a fixed ground field k.

A function is called a *constant* if, with the above notation, $f(x)$ is an element of k.

One might give another definition of functions, in the following manner. In the ring $\Omega[X]$ one takes the prime ideal $\mathfrak{P}$ associated with V. Denote by ξ_i the residue classes of the X_i mod $\mathfrak{P}$. Then $\Omega[\xi]$ is a ring, and its quotient field $\Omega(\xi)$ is canonically isomorphic to the function field of V. If k is a field of definition for V, then the mapping $(\xi) \to (x)$ gives an isomorphism of $k(\xi)$ onto $k(x)$.

Finally, note that one may also imbed the affine line S^1 canonically into the projective line, and that functions may be considered as rational maps of V into the projective line.

5. Correspondences

Rational maps, birational maps, and functions are special cases of what is known as a correspondence. We define a *correspondence* T between two varieties V and W to be a subvariety of the product $V \times W$. Two points (x') and (y') of V and W respectively are said to *correspond* under T if (x', y') is in T. If (x, y) is a generic point of T over a field of definition k for T, V, and W (in which case we say that T *is defined over* k) then (x) is a point of V and (y) is a point of W, but they are not necessarily generic points of V and W respectively. The set of points on W corresponding to (x) is the locus of (y) over $k(x)$, i.e. the set of all specializations (y') of (y) over $k(x)$. They form a $k(x)$-variety. All its components have therefore the same dimension, which is equal to the transcendence degree of (y) over $k(x)$.

If A is a closed subset of V, then we define $T(A)$ to be the projection on W of $T \cap (A \times W)$. If B is a closed subset of W, then we define $T^{-1}(B)$ to be the projection on V of $T \cap (V \times B)$. These are called the *image* of A, and *inverse image* of B respectively. Again, as with rational maps, a correspondence maps closed sets on closed sets. We write $T : V \to W$ by abuse of language (or more properly, by abuse of writing). It should be noted that if A is a point, then $T(A)$ coincides with the set theoretic image of A under T.

Let $T : V \to W$ be a correspondence between V and W, defined over k. Let (x, y) be a generic point of T over k, and let (x') be a point of V. If the coordinates y_j of (y) are integral over the local ring of (x') in $k(x)$, then we shall say that T is *finite over* (x'). We leave to the reader the verification that our definition is independent of the field of definition k, and the generic point (x) selected for T.

If T is finite over (x'), then any place of $k(x, y)$ inducing the specialization $(x) \to (x')$ must be finite on (y). (In fact, this condition could have been taken as the definition of T finite over (x'), according to Proposition 5 of Chapter I, § 5). Whenever a correspondence satisfies this condition, we say that it is *complete over* (x'). For our affine varieties, the two expressions "finite over (x')" and "complete over (x')" are synonymous, but this will not be the case for abstract varieties later.

If T is finite over (x'), then there is ony a finite number of points (y') on V corresponding to (x'). Indeed, the coordinates of (y') have to satisfy the specialized algebraic equations of the y_j over $(x) \to (x')$. The specialized equations are non-trivial by hypothesis. It is a consequence of Zariski's Main Theorem that the converse also holds (see the next chapter).

We shall say that T is of *degree n* if $k(x, y)$ is a finite extension of degree n over $k(x)$. We leave to the reader the verification that this is independent of the field of definition or generic point selected for T. If T is of degree n, then there are at most n distinct points in $T(x)$, which consist of the distinct conjugates of (y) over $k(x)$. If the above extension is separable, then of course there are exactly n points.

For the next theorem it is convenient to make a definition. Let V be a variety defined over a field k, (x) a generic point of V over k, and (x') a point of V. If the local ring of (x') in $k(x)$ is integrally closed, then we say that (x') is *k-normal*.

THEOREM 1. *Let $T : U \to V$ be a correspondence of finite degree d. Assume for convenience that the projection of T on U is U. Let k be a field of definition for T, U, V and let (x') be a point of U which is k-normal on U. Assume that T is finite over (x'). Then there are at most n points in $T(x')$. In fact, if (x, y) is a generic point of T over k, and if $(y^{(1)}, \ldots, y^{(d)})$ is a complete set of conjugates of (y)*

over $k(x)$, and $(\eta^{(1)}, \ldots, \eta^{(d)})$ is a specialization of this set over $(x) \to (x')$, then the points of $T(x')$ consists precisely of the points $(\eta^{(v)})$.

Proof: After raising the coordinates of (y) to a suitable power of the characteristic if necessary, we may assume that $k(x, y)$ is separable over $k(x)$. Let (u) be a set of m elements algebraically independent over $k(x, y)$ and $k(x')$. Put $K = k(u)$. Let $\mathfrak{o}$ be the local ring of (x') in $k(x)$ and let $R = \mathfrak{o}[u]$. It is easily proved that R is integrally closed (see below, the lemma of Proposition 9, Chapter V, § 4). Let $z = \sum u_j y_j$. Then z is integral over $R[x]$, and all the quantities $z_v = \sum_j u_j y_j^{(v)}$ are also integral over $R[x]$ and are distinct. Since $K(x, z)$ is contained in $K(x, y)$ these must be all the conjugates of z over $K(x)$. Let $f(Z)$ be the irreducible polynomial of z over K. It has leading coefficient 1, and its coefficients are in R by Proposition 9 of Chapter I, § 6. In addition, by Theorem 2 of Chapter I, § 4, the set $(z_1, \ldots, z_d)$ has a uniquely determined specialization $(\lambda_1, \ldots, \lambda_d)$ over $(x) \to (x')$, consisting of the roots of the reduced polynomial $\bar{f}(Z)$. This specialization can be extended to a specialization of $(y^{(1)}, \ldots, y^{(d)})$ because the y's are integral over $R(x)$. However, we have

$$f(Z) = \prod_v (Z - \sum_j u_j y_j^{(v)})$$

and hence

$$\bar{f}(Z) = \prod_v (Z - \sum_j u_j \eta_j^{(v)}).$$

Since (u) is a set of quantities which are algebraically independent over $k(x')$, and since the $(\eta^{(v)})$ must be algebraic over $k(x')$ it follows that the specialization $(\eta^{(1)}, \ldots, \eta^{(d)})$ is uniquely determined, and this proves our theorem.

We could have given an alternative proof of the proposition, first by reducing ourselves to the case where k is algebraically closed and (x') rational over k, and then using Theorem 3 of Chapter I, § 6. However, we considered the technique used above of sufficient interest to be given here.

As usual, we say that T is *finite over a subvariety A of V* if it is finite over a generic point of A (relative to a field of definition for

T and A). This property of T is independent of the field of definition selected. If T has the above property, then there is only a finite number of varieties A^* contained in T whose projection is A. This is seen immediately by looking at the generic points. Furthermore, there always exists such a subvariety. Indeed, if (x, y) is a generic point for T over a suitable field of definition k, and (x') is a generic point of A over k, then the specialization $(x) \to (x')$ can always be extended to a specialization (x', y') of (x, y), and (x', y') is the generic point of a k-subvariety of T containing A.

If we take k algebraically closed, then $k(x', y')$ is regular over k, and (x', y') is the generic point of a variety. Again looking at the integral equations satisfied by the y_j over the local ring of (x') in $k(x)$, we see that the y'_j satisfy an algebraic equation over $k(x')$. The variety A^* whose generic point is (x', y') can also be viewed as a correspondence, and we see that it is of finite degree, and finite over (x').

We shall now study correspondences which are not finite over a point. Our next result gives some information on the dimension of the algebraic set corresponding to such a point. It is a direct application of the dimension theorem of Chapter II, and will be called the *dimension theorem for correspondences*.

THEOREM 2. *Let $T \subset V \times W$ be a correspondence between V and W. Let A be a subvariety of V and let A^* be a component of $T \cap (A \times W)$ whose projection on the first factor is A. Let B be a subvariety of A, and let B^* be a component of $T \cap (B \times W)$ whose projection on the first factor is B. Assume that B^* is contained in A^*. Then*

$$\dim B^* - \dim B \geq \dim A^* - \dim A.$$

Proof: Notice first that we may assume $V = A$, $T = A^*$, and $W =$ the projection of A^* on the second factor. Without loss of generality, we deal therefore with a correspondence $T \subset V \times W$, a subvariety B of V, and we let B^* be a component of $T \cap (B \times W)$ having projection B on the first factor. We must show that $\dim B^* - \dim B \geq \dim T - \dim V$. Let $r = \dim V$. If $V = S^r$ is the full affine space, then we apply the dimension theorem.

If S^m is the affine space in which W lies, then intersecting T and $B \times S^m$ on $S^r \times S^m$, we see that the dimension of any component must be at least $\dim T + \dim B + m - (r + m) = \dim T + \dim B - r$, which is a reformulation of the result to be proved. (Note that any component of $T \cap (B \times S^m)$ will be contained in $B \times W$ because T is contained in $V \times W$.)

We shall now reduce our theorem to the case where $V = S^r$. Using the Noether normalization theorem, we can find an everywhere defined rational map $f : V \to S^r$ of V onto affine space, of finite degree, such that V is complete over every point of S^r, and such that to any point of S^r corresponds only a finite number of points of V. Let $f(B) = B_0$. It is a subvariety of S^r. As we are interested in local intersections at B, we may assume that $f^{-1}(B_0)$ has only one component, namely B itself. The map f induces a rational map $F : V \times W \to S^r \times W$. Let $B_0^* = F(B^*)$, and let $T_0 = F(T)$. Clearly, $F^{-1}(T_0) = T$. Let C_0^* be a component of $T_0 \cap (B_0 \times W)$ containing B_0^*. Then taking the inverse image back to $V \times W$, we see that

$$F^{-1}(T_0) \cap (f^{-1}(B_0) \times W) \supset C^* \supset B^*$$

for a suitable variety C^* in $F^{-1}(C_0^*)$. Since B^* is a component of $T \cap (B \times W)$, we must have $C^* = B^*$. The dimension of C^* is equal to that of C_0^*, and since we know the truth of our theorem in case $V = S^r$, we know that

$$\dim C_0^* - \dim B_0 \geq \dim T_0 - \dim S^r.$$

As we have trivially $\dim C_0^* = \dim C^*$, $\dim B = \dim B_0$, and $\dim T = \dim T_0$, we see that our theorem is proved.

We shall now give examples of correspondences which blow up subvarieties of a given variety V. They are due to Zariski, who calls them monoidal transformations.

Let (x) be a generic point for a variety V over a field k. Let W be a subvariety of V, defined over k, and let $y_1, \ldots, y_m$ be a basis for the ideal of W in $k[x]$. Let (x') be a generic point of W over k, and let φ be a place mapping (x) on (x'). Say y_m has maximal value among the y_i under the valuation associated with φ. Then $\varphi(y_i/y_m)$

is finite for all i, and if we put $z_i = y_i/y_m$ then we have a specialization $(x, z) \to (x', z')$ where $(z') = \varphi(z)$. Let U be the locus of (x, z) over k. Note that y_m is an element of the coordinate ring $k[x, z]$ of U. Intersect U with the hypersurface $y_m = 0$. Let W^* be a k-component containing (x', z'). Since y_m goes to zero, and the z_i are finite, all y_i must also go to zero. Hence the projection of W^* on V is equal to W. By the dimension theorem, dim $W^* = r - 1$ if $r = \dim V$. We see therefore that under the birational correspondence $T : V \to U$ such that $T(x) = (x, z)$, the subvariety W corresponds to an $r - 1$ dimensional subvariety of U.

Although we have not yet defined formally projective varieties and birational correspondences between them (this will be done in the section on abstract varieties), we nevertheless assume here that the reader is acquainted with the notion, and we state the following useful result.

Let V be a projective variety defined over a field k, and let A be a proper k-closed subset of V. Then there exists a birational correspondence $T : V \to U$ (also defined over k) between V and a projective variety U, such that:

1. T is defined at any point of V which does not lie in A and is not defined at any point of A.

2. T^{-1} is everywhere defined on U.

Proof: Let $(x_0, \ldots, x_n)$ be a homogeneous generic point of V over k, and let $R = k[x]$ be a homogeneous coordinate ring. By a form of degree d in R, we shall mean an element of R of type $f(x)$ where $f(X)$ is a form of degree d in the polynomial ring $k[X_0, \ldots, X_n]$. Let $g_0, \ldots, g_s$ be a basis for the homogeneous ideal $\mathfrak{a}$ in $k[x]$ associated with A. Let d_j be the degree of g_j, and let d be $> \max d_j$. If f is a form of degree d which lies in $\mathfrak{a}$, then we can write $f = h_0 g_0 + \ldots + h_s g_s$ where h_j is a form of degree $d - d_j$ in R, and conversely, every such expression is a form of degree d in $\mathfrak{a}$. Hence if $M_\alpha^{(j)}$ ranges over all monomials in (x) of degree $d - d_j$, then the $M_\alpha^{(j)} g_j$ generate the space of forms of degree d which lie in $\mathfrak{a}$. We denote this space by $\mathfrak{a}_d$. Any monomial $M(x)$ of degree ≥ 1 can be written $x_i M'(x)$ for some x_i, and another monomial $M'(x)$ of degree $d - 1$. Since $d > \max d_j$, we see that if $f_0, \ldots, f_N$ are in $\mathfrak{a}_d$ and generate $\mathfrak{a}_d$ as a vector space over k, then $x_i f_\mu (i = 0, \ldots, n$ and $\mu = 0, \ldots, N)$ generate $\mathfrak{a}_{d+1}$. We let V' be the projective variety having $(x_i f_\mu)$ as homogeneous generic point. Then we have a birational map $T : V \to V'$, because the field generated over k by all quotients $x_i f_\mu / x_j f_\lambda$ is clearly the same as that generated by all x_i/x_j. If (x') are the homogeneous coordinates of a point of V which is not in A, then T is defined at that point, because not all $g_0(x') = 0$, and hence not all $f_\mu(x') = 0$. However, T is not defined at any point A. Finally, T^{-1} is everywhere defined on V'. Indeed, let V_0 be the affine representative of V' having affine generic point say $(x_i f_\mu / x_0 f_0)$. Put $(\xi) = (x/x_0)$. Then this affine generic point is of type (ξ, ω) for some quantities (ω), and by definition, we see that T^{-1} is simply the projection, i.e., T^{-1} applied to the generic point having affine representative (ξ, ω) on V_0 is the point of V having affine representative (ξ). Hence T^{-1} is everywhere defined, and this concludes the proof of our proposition.

Later, we shall define normal varieties, and prove Zariski's Main Theorem concerning such varieties. It will be an immediate consequence of this that the subvarieties of A actually get blown up to higher dimensional subvarieties of U, just as in the affine case. One could also show this directly, but we leave it to the reader as an exercise.

6. Abstract varieties

There is at present a considerable amount of experimentation in the literature with the definition of abstract varieties. It is not our purpose here to experiment further, but rather to select one practical definition for certain simple applications, and to give the reader an idea of their use and purpose. Hence we take the definition given by Weil in *Foundations*. We shall show how certain theorems concerning abstract varieties are of a local nature, and are immediate consequences of the theorem for affine varieties. This is the reason why we have treated the theory of affine varieties fairly thoroughly. One should add that of course there is no substantial difficulty which arises in extending local theorems from affine to abstract varieties, but that there are sometimes unfortunately many tedious details to be carried out, not unlike those which arise in the foundations of the theory of differentiable manifolds. It is the custom to leave to the reader many of these details, and we shall not break with the custom here.

Let $V_\alpha (\alpha = 1, \ldots, h)$ be a finite number of affine varieties and suppose we have birational correspondences $T_\alpha^\beta : V_\beta \to V_\alpha$ which are coherent, i.e. such that $T_\gamma^\alpha T_\alpha^\beta = T_\gamma^\beta$. This means that if k is a field of definition for the V_α and the T_α^β then we can find generic points P_α of V_α over k such that all the fields $k(P_\alpha)$ are equal, and such that $T_\alpha^\beta(P_\beta) = P_\alpha$. In that case, (P_β, P_α) is a generic point of T_α^β over k.

In addition, suppose we are given a proper closed subset F_α on each V_α, and its complement U_α, satisfying the following condition: whenever two points $P'_\alpha \in U_\alpha$ and $P'_\beta \in U_\beta$ correspond under T_α^β, i.e. are such that $(P'_\beta, P'_\alpha) \in T_\alpha^\beta$ then T_α^β is biholomorphic at P'_β and P'_α.

Whenever this condition is satisfied, we say that $[V_\alpha, F_\alpha, T_\alpha^\beta]$ define an *abstract variety* V. If k is a field of definition for the V_α and the T_α^β and if each F_α is k-closed then we say that k is a *field of definition* for V. If K is a finitely generated regular extension of k, and can be written $k(P_\alpha)$ where P_α is the generic point of one of the V_α, then we say that V is a *model of* K and that K is *a function field for V over k.*

Let P_α $(\alpha = 1, \ldots, h)$ be corresponding generic points for the affine varieties V_α as above. Let φ be a place of K over k. Then φ may send some of the coordinates of the P_α to infinity. Suppose that for some index α none of the coordinates of P_α are mapped on infinity. Let $P_\alpha = (x_1, \ldots, x_n)$. Then $\varphi(P_\alpha) = (x'_1, \ldots, x'_n)$ is a point on V_α. If in addition P'_α is not in F_α or, equivalently, is in U_α, then we shall say that P'_α is a *representative* of an abstract point P' of V. The other representatives of P' are those points P'_β on the models V_β induced by the place, and not lying in F_β, i.e. which correspond to P'_α under T_α^β, the correspondence being biholomorphic by definition. We say that P' is a specialization of P.

If P' is a point of V, and P'_α, P'_β are two representatives, then the biholomorphy of T^β_α implies that $k(P'_\alpha) = k(P'_\beta)$. We may therefore denote this field by $k(P')$.

If, say, $P'_1, \ldots, P'_m$ is a maximal set of representatives for P', and if $k(P')$ is a regular extension of k, then each P'_α can be viewed as a generic point of a variety $V'_\alpha (\alpha = 1, \ldots, m)$. It is then clear that the V'_α can be used to define an abstract variety V', using the k-closed sets $F_\alpha \cap V'_\alpha = F'_\alpha$ and the birational correspondences T'^β_α having generic points (P'_β, P'_α). This abstract variety V' is called a subvariety of V, and is said to be the locus of P' over k.

On an abstract variety V we can define the Zariski topology as follows: A closed set is the finite union of subvarieties of V. To define a k-closed set, we have defined above the notion of specialization of a point. It is clear how to define the V^σ, where σ is an isomorphism of a field of definition for the abstract variety V, and V^σ, where σ is an automorphism of the universal domain. All the properties defining the notion of k-closed sets in Chapter III, § 3 are then meaningful, except of course the last two which refer to an imbedding in affine space. They are easily proved equivalent for an abstract variety, using Proposition 8 of § 3. Furthermore, using that same proposition we have the following statement which reduces the study of closed sets on an abstract variety to the study of closed sets on an affine representative.

PROPOSITION 13. *Let* $V = [V_\alpha, F_\alpha, T^\beta_\alpha]$ *be an abstract variety defined over a field* k. *Then the set of points of* V *which do not have a representative on a given* V_α *form a* k-*closed subset of* V.

The results concerning the smallest field of definition of an affine variety, and of rationality of cycles can be extended to abstract varieties by using the following criteria.

PROPOSITION 14. *Let* V *be as above an abstract variety defined over a field* k. *Let* W *be a subvariety of* V. *Then* W *has a smallest field of definition containing* k.

Proof: Let L be the compositum of k and of the smallest field of definition for any one of the representatives of W. Then clearly, W is defined over L. On the other hand, every field of definition for W must contain the smallest field of definition for each of its representatives, and if such a field contains k, it must contain L.

The free abelian group generated by the subvarieties of dimension s of V is called the group of *s-chains*. (Eventually we define simple points, and prove that they are invariant under biholomorphic transformations. Using this fact, one can define the notion of a simple point and simple subvariety on an abstract variety, and the free abelian group generated by the simple subvarieties of dimension s is called the group of *s-cycles on V*.)

Let W be a subvariety of an abstract variety V, defined over k. Suppose W is defined over a field $K \supset k$. Any isomorphism σ of K over k determines in an obvious manner a subvariety W^σ of V. If K is algebraic over k, and if Q is a generic point of W over K, then the order of inseparability of $k(Q)$ over k is defined to be that of W over k. We can then define the notion of a chain on V rational over k just as we did for affine varieties, and the following obvious proposition reduces all questions of rationality on an abstract variety to corresponding questions on an affine variety.

PROPOSITION 15. *Let Z be an s-chain on an abstract variety V defined over a field k. Let V_α be one of the affine representatives of V. Let Z_α be the chain on V_α which is obtained from Z by omitting all components which do not have a representative on V_α. Let K be a field containing k. Then Z is rational over K if and only if each Z_α is rational over K.*

Here of course, the chain Z_α is a cycle on the ambient affine space S_α of V_α. The results concerning smallest fields of rationality for cycles on affine space can then be immediately extended to abstract varieties by means of the proposition. The proposition according to which a subvariety of V has a smallest field of definition containing k appears as a special case of the theorem according to which a chain on V none of whose coefficients is divisible by the characteristic has a smallest field of rationality containing k. In fact, the results in affine space are special cases of the results for abstract varieties, because we can always view an affine variety V_α as a subvariety of S_α, and note that the prime field is a field of definition for S_α.

Products of abstract varieties can be defined in the following manner. Let $V = [V_\alpha, F_\alpha, T_\alpha^\beta]$ and $W = [W_\lambda, G_\lambda, R_\lambda^\nu]$ be two abstract varieties. Then $V \times W$ will be the abstract variety having affine representatives $V_\alpha \times W_\lambda$, and closed sets $(V_\alpha \times G_\lambda) \cup (F_\alpha \times W_\lambda)$. As for the birational correspondences, let (P_β, P_α) be a generic point of T_α^β and (Q_ν, Q_λ) a generic point of R_λ^ν over a common field of definition k for V and W. Then the birational transformation between $V_\alpha \times W_\lambda$ and $V_\beta \times W_\nu$ is the locus of $(P_\alpha, Q_\lambda, P_\beta, Q_\nu)$ over k. The verification that this is an abstract variety presents no difficulties.

One can then define correspondences and rational maps between abstract varieties in a manner similar to that for affine varieties.

However there is an alternative way of defining rational maps which we shall now describe. Let $V = [V_\alpha, F_\alpha, T_\alpha^\beta]$ and $W = [W_\gamma, G_\gamma, R_\lambda^\nu]$ be two abstract varieties defined over a field k. Let A be the set of indices $\{\alpha\}$ and C the set of indices $\{\gamma\}$. A rational map $f : V \to W$ consists of a collection of rational maps $f_\gamma^\alpha : V_\alpha \to W_\gamma$, where α ranges over all indices in A, and γ ranges over a subset C_0 of C, satisfying the following property. If P_α is a generic point of V_α over a common field of definition for V, W, and all f_α then the points $Q_\gamma = f_\gamma^\alpha (P_\alpha)$ form a complete set of representatives for a point Q of W. One then writes $f(P) = Q$. Since $k(Q) \subset k(P)$, the field $k(Q)$ is a regular extension of k, and Q is the generic point of a subvariety U of W. We write $f(V) = U$. If $W = U$, then we say that f is generically surjective. If k is a field of definition for V, W, and all the f_γ^α, we say f is defined over k.

Conversely, it is clear that if V and W are two abstract varieties defined over k, U is a subvariety of W also defined over k, P is a generic point of V over k, and Q a generic point of U over k such that $k(Q) \subset k(P)$, then there is a unique rational map $f : V \to W$ such that $f(P) = Q$. In terms of a fixed function field K for V, a rational map consists therefore in giving a model of a subfield of K.

If P' is a point of V, then all the local rings in K of its affine representatives are equal by definition, so we can speak of the local ring of P'. If Q' is a point of U, then it has a local ring in the subfield $k(Q)$. We say that f is defined at P' if at least one of the rational maps f_γ^α is defined at P'_γ. This is easily seen to be equivalent to the following invariant statement: The canonical homomorphism of the local ring of P' in $k(P)$ is at the center of the canonical homomorphism of the local ring of Q' in $k(Q)$. We say f is defined at a subvariety if it is defined at a generic point of that subvariety.

Correspondences can be treated in the same manner. For instance, if P and Q are two points of abstract varieties V and W respectively, such that $k(P, Q)$ is a regular extension, then there is a unique correspondence $T : V \to W$ such that $T(P)$ is equal to the locus of Q over $k(P)$. It is a $k(P)$-closed subset of W. One says also that (P, Q) is a generic point for the correspondence.

A rational map $f : V \to W$ is said to be birational if it is generically surjective, and if $k(P) = k(Q)$ in the notation used above. More functorially, one may say that f is birational if there exists a rational map $g : W \to V$ such that g is defined at $f(V)$ and such that $g \circ f = $ identity. One then says that f and g are inverse to each other. Suppose this is the case. Let P' and Q' be two points of V and W respectively. We say that f and g are biholomorphic at P' and Q' if f is defined at P', $f(P') = Q'$, g is defined at Q' and $g(Q') = P'$. By abuse of language, we shall also say that f is biholomorphic at P'. If f is biholomorphic at every point, we say that f is an isomorphism between V and W. If f is defined over k, we say that f is a k-isomorphism. Similar definitions can be made for subvarieties V' and W' of V and W, instead of points P' and Q'.

The rational maps of V into S^1 form a field $\Omega(V)$ exactly as for affine varieties. If P is a generic point of V over k, and f is such a rational map (which is called a function on V), then $f(P)$ is an element of $k(P)$. The field $k(V)$ of all functions on V defined over k is isomorphic to $k(P)$ under the mapping $f \to f(P)$.

Let now $f : V \to W$ be an isomorphism. Then there is a $1 - 1$ correspondence between the subvarieties of V and W. If k is a field of definition for $f, V,$ and W this correspondence is established as follows. Let V' be a subvariety of V. Suppose V' is defined over k. Let P' be a generic point of V' over k and put $f(P') = Q'$. Then $k(P') = k(Q')$, and Q' is the generic point of a subvariety W' of W, which is also defined over k. If g is the inverse of f, then we see that $g(W') = V'$. We can obviously extend f and g to isomorphisms of the groups of chains on V and W respectively and we get

PROPOSITION 16. *Let $f : V \to W$ be a k-isomorphism between two abstract varieties V and W. Let Z be a chain on V and $f(Z)$ the corresponding chain on W. Then Z is rational over a field K containing k if and only if $f(Z)$ is rational over K.*

In particular, we see that the smallest field of rationality of a chain, if it exists, is a biholomorphic invariant.

All local theorems concerning rational maps and correspondences on affine varieties can be extended without difficulty to abstract varieties. The following auxiliary result can sometimes be used to make the connection between affine and abstract varieties technically easier.

PROPOSITION 17. *Let V be an abstract variety defined over a field k. Then there exists a biholomorphic correspondence $f : V \to W$ defined over k, between V and an abstract variety $W = [W_\gamma, G_\gamma, R_\lambda^\gamma]$, such that all frontiers G_γ are empty.*

Proof: Given an affine representative V_α of V, let $\mathfrak{a}$ be the ideal in a coordinate ring $k[x]$ of V_α vanishing on the frontier F_α of V_α. Let $z_1, \ldots, z_m$ be a basis for $\mathfrak{a}$, such that all $z_j \neq 0$. Consider the affine varieties $V_{\alpha j}$ which are the locus of $(x, 1/z_j)$ over k. Then it is clear that the $V_{\alpha j}$, empty frontiers and the obvious birational correspondences, define the abstract variety W which we are looking for.

An abstract variety is said to be *complete* if the following condition is satisfied. Let P be a generic point of V over k. Let φ be a place of $k(P)$ over k. Then for at least one affine representative P_α, $\varphi(P_\alpha)$ is a point of V_α (i.e. all coordinates are finite under the place) and does not lie in F_α. This condition is easily seen to be independent of the field of definition selected. The hypothesis that a variety is complete is used, among other things, to guarantee that certain intersections of subvarieties are not empty, or that the image of a closed set under a correspondence is not empty. One can then study intersections and correspondences or rational maps on affine models. In other words, there is a certain type of question for which the global condition of completeness contributes nothing except the knowledge that a certain variety has not escaped to infinity.

Let T be a correspondence between two abstract varieties V and W. Suppose that T, V, and W are defined over k. In terms of one given field, this means that we have a point P of V, and a point Q of W, such that (P, Q) is viewed as a generic point of T over k, the extension $k(P, Q)$ over k being regular. We shall say that T is *complete over a point P' of V*, if every place of $k(P, Q)$ inducing the specialization $\varphi(P) = P'$ is such that it also induces a point Q' of W, i.e. for some representative Q_λ of Q, $\varphi(Q_\lambda)$ is a point on W_λ not in G_λ. If W is complete, then of course T is complete over every point of V. We say that T is *complete over a subvariety of V* if it is complete over a generic point of that subvariety. This situation is considerably different from that for affine varieties, since in that case T may well be complete over a point, and yet a whole subvariety of W may correspond to that point.

We shall say that T is *finite* over a point P' of V if T is complete over P', and there is only a finite number of points Q' on W corresponding to Q. Zariski's Main Theorem will show how this definition is related with the integral closure of the local ring of P' in the field $k(P, Q)$.

Returning to the notion of completeness, we see that if we view a correspondence T between V and W as a subvariety of the product, and if W is complete, then the projection of T on V coincides with set theoretic projection. Indeed, if (P, Q) is a generic point of T over some field k, then P is a generic point of its projection on V. If P' is a point of the projection, i.e. a specialization of P over k, then we can extend this to a place of $k(P, Q)$, which induces a point Q' on W, and (P', Q') correspond under T. In particular, we get

PROPOSITION 18. *If $f : V \to W$ is a rational map of a complete variety V into an abstract variety W, and is everywhere defined, then $f(V)$ is a complete subvariety of W, and f maps V set theoretically onto $f(V)$.*

We shall now study projective varieties, and we shall see that if V is an abstract, complete variety, then there is a projective variety U and a birational map $f : U \to V$ everywhere defined on U, and mapping U onto V.

Projective spaces are special cases of complete abstract varieties. Indeed, let $u_0, \ldots, u_n$ be algebraically independent quantities over a field k, and let V_α be the affine variety having $(u_0/u_\alpha, \ldots, u_n/u_\alpha) = P_\alpha$ as generic point. Then V_α, empty closed sets, and the obvious birational correspondences whose generic points are $(P_\beta, P_\alpha)(\alpha, \beta = 0, \ldots, n)$ define what is called projective space. A subvariety of projective space is a projective variety. If φ is a place of $k(u_0, \ldots, u_n)$ over k, and if $|u_\alpha| \geqq |u_j|$ for all $j = 0, \ldots, n$ then φ induces a point on the affine representative V_α, so projective space is complete.

It is trivially verified that a subvariety of a complete variety is also complete. Hence projective varieties are complete.

In many theories, concerned with projective varieties, no use is ever made of the projective properties of the varieties except their completeness. In such theories, it is therefore much clearer to deal with abstract complete varieties, until such time as a projective property is really needed. In our chapter on divisors and linear systems, the reader will actually see such a situation where the projective varieties are used in an essential way.

If V is a projective variety, defined over a field k, then the following is easily verified. There is a unique homogeneous variety V^* such that if $(x_0, \ldots, x_n)$ is a homogeneous generic point of V^* over k, and if we put $\xi_i = x_i/x_\alpha$ for some α such that $x_\alpha \neq 0$, then (ξ) is the affine representative on V_α of the projective variety V, and all such affine representatives are obtained in that way. Conversely, every homogeneous variety V^* in $n + 1$ affine space gives rise to a projective variety. A homogeneous generic point of V^* will also be called a *homogeneous generic* *point of* V, or a set of *homogeneous coordinates* for that point.

The homogeneous ideals in the full polynomial ring $k[X]$ discussed in Chapter II can of course be generalized to homogeneous ideals in a coordinate ring $k[x]$, where (x) is a homogeneous generic point. By a form in $k[x]$ we mean an element of $k[x]$ of type $f(x)$ where $f(X)$ is a form in $k[X]$. To each k-closed subset A of V, it is easily seen that we can associate a unique homogeneous ideal $\mathfrak{a}$ of $k[x]$, consisting of all forms $f(x)$ in R satisfying $f(x') = 0$ for any set of homogeneous coordinates (x') belonging to a point of A. Conversely, every homogeneous ideal in $k[x]$ defines a k-closed subset of V, consisting of all points of V whose homogeneous coordinates are a zero of $\mathfrak{a}$.

The product of two projective varieties has a projective embedding. Indeed, Let V and W be two projective varieties, and let (x) and (y) be two homogeneous generic points of V and W over a common field of definition. Then the point $(x_i y_j)$, where x_i ranges over all coordinates of (x), and y_j over all those of (y), is a homogeneous generic point of some projective variety U, and the reader will immediately verify that U is biholomorphic to $V \times W$.

The following lemma of Chow can frequently be used as a substitute for the projective embedding of an abstract variety.

PROPOSITION 19. *Let V be an abstract variety, defined over a field k. Then there is a birational correspondence $T : U \to V$ between a projective variety U and V, defined over k, and there exists a k-closed subset A of U, such that:*

1. *T is everywhere defined on $U - A$ and maps $U - A$ onto V;*

2. *no point of A corresponds to a point of V under T; if V is complete, then A is empty.*

Proof: The abstract variety V is given as $[V_\alpha, F_\alpha, T_\alpha^\beta]$. Actually, we may take each V_α to be a projective variety, and extend F_α so as to include the hyperplane at infinity with respect to the given affine variety. The T_α^β are then extended in the same fashion. Let $P_1, \ldots, P_m$ be a set of corresponding generic points of the V_α, so $k(P_\alpha) = k(P_1, \ldots, P_m)$ for any α. We may view $(P_1, \ldots, P_m)$

as the generic point of a projective variety U in the following manner. Say $m = 2$, and $(x_0, \ldots, x_r)$, $(y_0, \ldots, y_s)$ is a set of homogeneous coordinates for P_1, P_2. Then $(x_i y_j)$ is a set of homogeneous coordinates for a point Q in some projective space, whose locus over k is biholomorphic to that of (P_1, P_2) in the product space $V_1 \times V_2$. This is extended inductively to the m points $P_1, \ldots, P_m$. We let U be the locus of $(P_1, \ldots, P_m)$ over k. Then U is projective, and there is a birational map $T \cdot : U \to V$ mapping $(P_1, \ldots, P_m)$ on the point of V having these P_α as its representatives. For each α, we have a birational map $\pi_\alpha : U \to V_\alpha$, namely the projection. If $Q' = (P'_1, \ldots, P'_m)$ is a point of U, then $\pi_\alpha(Q') = P'_\alpha$. We see that π_α is everywhere defined. Hence $\pi_\alpha^{-1}(F_\alpha)$ is a k-closed subset S_α of U. We let S be the intersection of all the S_α. Then S is a k-closed subset of U. If Q' is a point of U which is not in S, then for some α, $\pi_\alpha(Q')$ is not in F_α, and hence the birational map $T : U \to V$ maps $U - S$ into V. It is clear that T does in fact map $U - S$ onto V. Furthermore, no point of S corresponds to a point of V under T, for if Q' is in S, then each P'_α is in F_α. If V is complete, then S must be empty, because a place of the function field $k(P_\alpha)$ inducing a point Q' in S on U cannot induce any point on V.

Literature

As we have said earlier, this chapter consists mostly of definitions. The only part which merits comment is again the part dealing with abstract varieties. These have become of considerable importance in algebraic geometry.

In the first place, they are the natural tool to construct fibre spaces [1] [7]. Furthermore, they are also used by Weil to construct transformation spaces and homogeneous spaces [8] [9] and by Serre [6] in his cohomology theory of sheaves. Nagata has shown that a complete non-singular variety may not have a projective embedding if its dimension is ≥ 3. (This is in course of publication.) On the other hand, Chow [3] has proved that all homogeneous spaces can be so embedded. This problem is closely connected with linear systems, as we shall see in a subsequent chapter.

For other definitions of an abstract variety, we refer the reader to Serre [6] and for the arithmetic point of view, to Chevalley [2] and Nagata [4].

CHERN, S. S.

[1] On the characteristic classes of complex sphere bundles and algebraic varieties, Am. J. of Math., Vol. 75, No. 3, July 1953, pp. 565—597.

CHEVALLEY, C.

[2] Sur la théorie des variétes algébriques, Nagoya Math. J., Vol. 8, February 1955, pp. 1—43.

CHOW, W. L.

[3] Projective embedding of homogeneous spaces, Lefschetz conference collection of papers, Princeton. 1957.

NAGATA, M.

[4] A general theory of algebraic geometry over Dedekind domains, Am. J. of Math., Vol. 78, No. 1, January 1956, pp. 78—116.

[5] On the embedding problem of abstract varieties in projective varieties, Memoirs of the college of science, University of Kyoto, Vol. XXX, No. 1, 1956, pp. 71—82.

SERRE, J. P.

[6] Faisceaux algébriques cohérents, Annals of Math., Vol. 61, No. 2, March 1955, pp. 197—278.

WEIL, A.

[7] Fibre spaces in algebraic geometry, Notes by A. Wallace, University of Chicago, 1952.

[8] Algebraic groups of transformations, Am. J. of Math., Vol. 77, No. 2, April 1955, pp. 355—391.

[9] On algebraic groups and homogeneous spaces, Am. J. of Math., Vol. 77, No. 3, July 1955, pp. 493—512.

CHAPTER V

Normal Varieties

We give some useful results relative to the integral closure of a ring in its quotient field and in a finite extension, and the applications of these results to varieties. We treat affine varieties first, as usual, and then abstract and projective varieties. We also include Zariski's original proof of his Main Theorem on birational correspondences, but the section dealing with this theorem is not used anywhere else in this book except to prove some uniqueness statements concerning the normalization of an abstract variety.

Next, we present the more detailed theory dealing with projective normalization. This is essentially a projective theory, depending strongly on the embedding of the variety in projective space. The results are due to Zariski, and include some unpublished results of his, reproduced here with his permission.

Finally, we deal with the problem of constant field extensions, and prove that a point which is k-normal remains normal under separable extension of the ground field, and that the product of two k-normal points on the product variety is also k-normal. This is the only section of this chapter which depends on Chapter III.

1. Integral closure of an affine ring

We recall that a ring R is said to be integrally closed if it is integrally closed in its quotient field K. An element y of a finite extension L of K is integral over R if it satisfies an integral equation, or if every place finite of L finite on R is finite on y. If y is an arbitrary element of L, then there exists an element c of R such that cy is integral over R. This is easily seen by writing the equation for y

$$c_n y^n + \ldots + c_0 = 0$$

with $c_j \in R$. Multiplying by c_n^{n-1}, we see that $(c_n y)$ is integral over R. Recall also that if y is integral over R, so is every conjugate of

y over K (apply an isomorphism leaving K fixed to an integral equation for y over R).

THEOREM 1. *Let R be a Noetherian ring, integrally closed in its quotient field K. Let L be a finite separable algebraic extension of K. Let I be the integral closure of R in L. Then there exist elements $y_1, \ldots, y_m$ in I such that $I = Ry_1 + \ldots + Ry_m$.*

Proof: Let $w_1, \ldots, w_m$ be a basis of L over K. After multiplying each w_i by a suitable element of R, we may assume that each w_i is integral over R. Let $w_1', \ldots, w_m'$ be the dual basis with respect to the trace of L over K. For any element x of I, we can write $x = a_1 w_1 + \ldots + a_m w_m$, with $a_i \epsilon K$, and we have $a_i = Tr(xw_i')$. There exists an element $b \epsilon R$ such that bw_i' is integral over R for all i, and hence $ba_i = Tr(xbw_i')$ is integral over R and therefore is in R because R is integrally closed. This proves that $bI \subset Rw_1 + \ldots + Rw_m$ and that $I \subset Rz_1 + \ldots + Rz_m$, where $z_i = w_i/b$. Since R is Noetherian, this implies that I is a finite R-module, as was to be shown.

The conclusion of the above theorem may be false if L is inseparable over K. However, the following result shows that it remains true in a special case important for algebraic geometry.

THEOREM 2. *Let $k[x] = R$ be a finitely generated extension of a field k, let K be the quotient field $k(x)$, and let L be a finite algebraic extension of K. Let I be the integral closure of R in L. Then I is a finite R-module.*

Proof: By the Noether normalization theorem, we can find $y_1, \ldots, y_r$ algebraically independent over k, and lying in $k[x]$ such that $k[x]$ is integral over $k[y]$. The integral closure of $k[y]$ in L is the same as that of $k[x]$ in L by transitivity, so we may assume that our elements (x) are algebraically independent over k. Furthermore, it suffices to prove the theorem for any finite extension of L, because a submodule of a finitely generated module over a Noetherian ring is finitely generated. In particular, we may assume that L is normal over K (but not necessarily separable). In that case $L \supset L' \supset K$, where L' is purely inseparable over K and L is separable over L'. By using the fact that if the (x)

are algebraically independent over k then $k[x]$ is integrally closed in K, and Theorem 1, we see that our theorem is reduced to the case of a purely inseparable extension of K.

Going up inductively, we may assume that L is generated by one element, and in fact $L = K(f(x)^{1/p})$ where p is the characteristic, and $f(x) \epsilon k[x]$. Let $\{c_j\}$ range over the finite number of coefficients of $f(x)$ in k. Then

$$k[x_1^{1/p}, \ldots, x_r^{1/p}, c^{1/p}] = k(c^{1/p})[x_1^{1/p}, \ldots, x_r^{1/p}]$$

and this ring is just a polynomial ring with coefficients in $k(c^{1/p})$. It is therefore integrally closed. We have

$$R \subset I \subset k(c^{1/p})[x_1^{1/p}, \ldots, x_r^{1/p}]$$

and this latter ring is integral over R, and is a finite module over R. Hence by the squeezing argument, I is also a finite module over R, as desired.

In geometric terms, taking $(x_1, \ldots, x_n) = (x)$ to be the generic point of a k-variety V, we let $y_1, \ldots y_m$ be a set of generators for the integral closure I of $k[x]$ in L (viewed as a module over $k[x]$). Then the k-variety U having (x, y) as a generic point over k is called the *k-normalization of V in L*.

For the rest of this chapter, we shall use varieties instead of k-varieties and hence will assume that both extensions $k(x)$ and L are regular over k. Then V and U are varieties. The reader will note, however, that we could continue with k-varieties throughout. Only rational maps and correspondences would require some discussion, because the product of two k-varieties is not always a k-variety. If the reader assumes k algebraically closed, then this trouble does not arise, and nothing is needed from Chapter III.

As we know, we may call L a finite extension of the function field $k(x)$, and U will also be called the normalization of V in a finite extension of its function field.

It is immediately seen that U is uniquely determined up to a k-isomorphism. Furthermore, the rational map $f : U \to V$ such that $f(x, y) = (x)$ has three fundamental properties:

1. It is everywhere defined on U.
2. Given a point (x') on V, there is only a finite number of

points on U mapping on (x'), so f is finite over every point of V.

   3. f is complete over every point of V.

   Let (x') be a point on the affine variety V defined over k, and let (x) be a generic point of V over k. We shall say that (x') is *k-normal* if its local ring in $k(x)$ is integrally closed. The variety V is said to be *k-normal at* (x'). If V is k-normal at every point, then it is said to be *k-normal*. If a point of V is k-normal for every field of definition k of V, it is said to be *normal*, and V is said to be *normal* if every point of V is normal. We make similar definitions for a subvariety of V provided its generic point has the corresponding property.

   The coordinate ring $R = k[x]$ being integrally closed, it will now be proved that every local ring is also integrally closed. Hence to get a k-normal variety, closing its coordinate ring suffices.

   PROPOSITION 1. *Let R be a ring. Then the local ring $R_{\mathfrak{p}}$ is integrally closed for every prime ideal $\mathfrak{p}$ of R if and only if R is integrally closed.*

   Proof: Assume first R integrally closed. Let z be an element of its quotient field K, integral over $R_{\mathfrak{p}}$, so we can write

$$z^m + (a_{m-1}/b_{m-1})z^{m-1} + \ldots + a_0/b_0 = 0$$

with a_i, $b_i \in R$, $b_i \notin \mathfrak{p}$. Multiply by the greatest denominator c equal to the product of the b's, and then multiply by c^{m-1}. Then we see that (cz) is integral over R, hence in R. We can therefore write $cz = d$, with $d \in R$. But $c \notin \mathfrak{p}$, as one sees from the construction of c, and hence $z \in R_{\mathfrak{p}}$, as was to be shown.

   Conversely, let z be in K, integral over R. Then z is *a fortiori* integral over each $R_{\mathfrak{p}}$, and hence in each $R_{\mathfrak{p}}$ by hypothesis. It is therefore in R (See Theorem 6 of Chapter II, § 3).

   Normal varieties are important for at least two reasons:

   1. Let V^r be defined over a perfect field k, and be normal. Let (x') be a point on V of dimension $r - 1$ over k. Its local ring is integrally closed, and it will be proved in Chapter VI that it is a discrete valuation ring. It will be shown in the chapter on simple points that such a point is simple. Hence we shall see that all

subvarieties of codimension 1 are simple on a normal variety.

2.   The second reason is Zariski's Main Theorem on birational correspondences, which we shall abbreviate ZMT. This theorem will be proved in the next section.

We shall now make a few more comments concerning the birational correspondence between a variety V and its normalization. Let us first prove a lemma.

LEMMA.   *Let R be a ring contained in a ring S integral over R and finitely generated (as a ring) over R, i.e. $S = R[y_1, \ldots, y_n]$. Then S is a finite R-module, i.e. there exist elements $w_1, \ldots, w_N$ in such S that $S = Rw_1 + \ldots + Rw_N$.*

Proof: If y is an element of S, then y satisfies an equation

$$y^m + a_{m-1}y^{m-1} + \ldots + a_0 = 0$$

with $a_i \in R$. All powers of y are then contained in the R-module $R + Ry + Ry^2 + \ldots + Ry^{m-1}$. If n_j is the degree of an equation of the above type for each y_j, then S is equal to the finite R-module generated by the monomials

$$y_1^{v_1} \ldots y_n^{v_n},$$

where v_j are integers, $0 \leq v_j < n_j$.

Let R be a ring contained in a ring S integral over R. The *conductor* of S over R is defined to be the set of elements $f \in R$ such that $fS \subset R$. One verifies immediately that the conductor is an ideal of both rings R and S, and in fact is the largest ideal of S contained in R.

PROPOSITION 2.   *Let $T : U \to V$ be a birational correspondence defined over a field k. Let (x) be a generic point of U over k, and $T(x) = (y)$. Assume that $k[x] = R$ is contained in $k[y] = S$ and that $k[y]$ is integral over $k[x]$. Then the set of points on U or V where T is not biholomorphic is the k-closed set of zeros of the conductor $\mathfrak{a}$ of S over R.*

Proof: Our statement depends on the fact that $\mathfrak{a}$ is an ideal of S and R, and hence determines an algebraic set both on U and V.

Let us begin with U. Let (x') be a point of U where T is not defined. If (x') is not a zero of the conductor, then there exists

$f(x)$ in a such that $f(x') \neq 0$. By definition, for each coordinate y_j of (y) we have $f(x)y_j = g_j(x)$. This shows that $y_j = g_j(x)/f(x)$, and that T is defined at (x'), a contradiction.

Conversely, suppose that T is defined at (x'). By the lemma, we can write $S = Rw_1 + \ldots + Rw_m$, and by hypothesis $w_i = f_i(x)/g_i(x)$, where $g_i(x') \neq 0$. Hence $g_i(x)w_i = f_i(x)$, and by definition, we see that the product g of the $g_i(x)$ must be in the conductor. But $g(x') \neq 0$, and hence (x') cannot be a zero of the conductor.

Let us now go over to V. Let (y') be a point of V where T^{-1} is not biholomorphic. If we put $T^{-1}(y') = (x')$, then T cannot be defined at (x'). If (y') is not a zero of the conductor (viewed as an ideal of $k[y]$) then (x') cannot be a zero of the conductor (viewed as an ideal of $k[x]$) and we have a contradiction as before. The converse is proved the same way.

COROLLARY. *Let U be a variety defined over a field k. The set of points on U where U is not k-normal is a proper k-closed subset of U.*

Proof: We apply the preceding proposition to the case where V is the k-normalization of U.

REMARK. It will be proved that if k is perfect, then a point is normal if and only if it is k-normal. Hence we see that the set of points where V is not normal is also a closed subset of V.

2. Zariski's Main Theorem (ZMT)

The theorem runs as follows. *Let $T : U \to V$ be a rational map defined over a field k. Let P be a point of U, which is k-normal and assume that $T \cap (P \times V)$ contains a component which is reduced to a point. Then this is the only component, and T is defined at P.*

We first note that P may be viewed as the generic point of a k-subvariety W of U. Our assumption then implies that $T \cap (W \times V)$ has a k-component of the same dimension as W and projecting on W, i.e. lying above W. Indeed, if (P, Q) is a component of the intersection of $T \cap (P \times V)$, then Q must be algebraic over $k(P)$, and (P, Q) may be viewed as the generic point over k of a k-variety W^*. One sees immediately that W^* must be itself a k-component of $T \cap (W \times V)$.

The proof of ZMT will take place in two steps. First we reduce the theorem to a special case. This reduction is essentially trivial. Secondly, we make a close analysis of this special case. This embodies the heart of the proof.

Reduction to a special case

Since the theorem is local with respect to P on U, we may replace U by any other variety biholomorphic to it at P. Hence we may assume that U is k-normal (and not only k-normal at P). If (x) is a generic point of U over k, then the coordinate ring $R = k[x]$ is integrally closed.

Furthermore, we may clearly replace V by T. We do this for convenience, so that we may assume that T is a birational map, and that T^{-1} is everywhere defined. Under this assumption, if (x) is a generic point of U over k and $T(x) = (y)$, then $k[x] \subset k[y]$. The hypothesis that $T(P)$ has a component reduced to a point still holds.

Finally, we may assume that V is k-normal, for it is an immediate consequence of the definition of normalization that if we replace V by its k-normalization V^*, then the properties listed above for the correspondence between U and V remain valid for the correspondence T^* between U and V^*. If T^* can be proved to be defined at P, then *a fortiori* T must also be defined at P.

We could dispense with the above trivial changes on U and V, but they will allow us slightly more freedom of language than we would otherwise have had.

We shall now interpose a sequence of varieties between U and V and show that it suffices to prove ZMT for each step of the sequence. According to the above remarks, if (x) is a generic point of U over k, and if we put $R = k[x]$, and $T(x) = (y)$, then $S = k[y]$ can be written $R[y_1, \ldots, y_m]$. We let R_i be the integral closure of $R_{i-1}[y_i]$, starting the induction with $R_0 = R$. In other words, at each step we adjoin one element and take the integral closure. Each R_i is the coordinate ring of a variety V_i and we have a birational correspondence $T_i : V_i \to V_{i+1}$ associated with the inclusion of the coordinate rings. Each T_i^{-1} is everywhere defined, and the point $Q = Q_m$ on V determines points Q_i on V_i such that $Q_{i-1} = T_{i-1}^{-1}(Q_i)$.

We consider our sequence from righ to left, starting with T_{m-1}. In the image of Q_{m-1} under T_{m-1} we certainly have $Q_m = Q$. If Q were not a component of $T_{m-1}(Q_{m-1})$ then one sees immediately that Q would not be a component in the image of P under T, contrary to assumption. Hence T_{m-1} satisfies the hypothesis of ZMT. If we have proved ZMT in that case, we can conclude that T_{m-1} is biholomorphic at Q_{m-1} and Q. It suffices therefore to prove ZMT for the transformation from U to V_{m-1}. Proceeding inductively, we see therefore that we have reduced the proof of ZMT to the case where T satisfies in addition the following hypotheses.

Special case. $T : U \to V$ is a birational correspondence, and if (x) is a generic point of U over k, $T(x) = (y)$, and $R = k[x]$, $S = k[y]$, then S is integral over the ring $R[y_1]$ obtained by adjoining one coordinate to R.

Proof in the special case: Let U_1 be the locus of (x, y_1) over k, and let $R_1 = k[x, y_1]$ be its coordinate ring. Then we have two birational correspondences

$$T_1 : U \to U_1 \quad \text{and} \quad T_2 : U_1 \to V.$$

Their inverses are everywhere defined by construction, and in particular, the point Q on V determines a point P_1 on U_1. Our proof will consist in proving first that T_2 may be assumed biholomorphic between P_1 and Q, and then proving ZMT for the transformation T_1 which is distinguished by the fact that the coordinate ring of U_1 is obtained by adjoining only one element to that of U.

We contend that we may assume in addition that T is biholomorphic at every point Q^* on V such that $\dim_k(Q^*) > \dim_k(Q)$ and such that $Q^* \to Q$ is a specialization over k. Our contention may also be expressed by saying that we may assume T^{-1} biholomorphic at every k-subvariety of V properly containing Q, because Q^* may be viewed as the generic point of such a subvariety. Indeed, given such a point Q^*, let $P^* = T^{-1}(Q^*)$. Then Q^* is in the image $T(P^*)$. It is in fact a component of this image, as one sees immediately from the dimension theorem for correspondences given in Chapter IV. The hypotheses of ZMT are therefore satisfied with respect to P^* and Q^*. If T is not biholomorphic at Q, we take Q^* of maximal dimension such that T is not biholomorphic at Q^* and such that Q is a specialization of Q^* over k. This proves our contention.

We now have the following proposition, which gives the biholomorphy of T_2 at P_1 and Q, and which can be regarded as the heart of the proof of ZMT. It is of considerable importance in itself.

PROPOSITION 3. *Let $T : U \to V$ be a birational correspondence defined over k, which can be decomposed into two steps,*

$$T_1 : U \to V_1 \quad and \quad T_2 : V_1 \to V$$

satisfying the following properties. If (x) is a generic point of U over k, then $T_1(x) = (x, y_1)$ is obtained by adding one coordinate, and $T_2(x, y_1) = (x, y)$ is integral over $k[x, y_1] = R_1$. Assume that $k[x] = R$ is integrally closed. Let $P, P_1,$ and Q be three corresponding points such that P_1 and Q are algebraic over $k(P)$, and assume that T_2 is biholomorphic at every k-subvariety of V properly containing Q. Then T_2 is biholomorphic at P_1 and Q.

Proof: The locus of Q over k is a k-variety which we denote by W. Suppose that T_2 is not biholomorphic at W (i.e. at Q by definition). Let $\mathfrak{a}$ be the conductor of $S = k[x, y]$ over $R = k[x]$. Its zeros on V form a k-closed set which can be expressed as a union $W \cup B$, where B is k-closed and does not contain W. This follows from Proposition 2 of § 1, the decomposition of algebraic sets into k-varieties, and the hypothesis which guarantees that W is a k-component. Then there exists an element t of S vanishing on B but not on W. Let $\mathfrak{p}, \mathfrak{p}_1,$ and $\mathfrak{q}$ be the associated prime ideals of $P, P_1,$ and Q respectively in the respective coordinate rings $R, R_1,$ and S. By the Hilbert Nullstellensatz, $(t\mathfrak{q})^N$ is contained in $\mathfrak{a}$ for a suitable power N. Furthermore, $t^N \notin \mathfrak{a}$ because t^N does not vanish on W. Hence there exists an element $w \in S$, such that

(1) $t^N w \notin R_1$ but $t^N q^N w \subset R_1$.

By hypothesis, the point P_1 is algebraic over $k(P)$, and hence after having localized if necessary, we have an equation

$$y_1^n + a_{n-1} y_1^{n-1} + \ldots + a_0 \in \mathfrak{p}_1$$

with coefficients a_i in R. Raising both sides to the Nth power and noting that $\mathfrak{p}_1 \subset \mathfrak{q}$, we get

$$G(y_1) \in \mathfrak{q}^N$$

where G is a polynomial with coefficients in R, and leading coefficient 1. Put $z = t^N w$. Then by (1) we have

$$z \notin R_1 \quad \text{and} \quad zG(y_1) = F(y_1)$$

for some polynomial F with coefficients in R. Since G has leading coefficient 1, we use long division to get

$$z = h(y_1) + \frac{g(y_1)}{G(y_1)},$$

where g and h are polynomials with coefficients in R, and deg $g <$ deg G. Put $u = g(y_1)/G(y_1)$. We shall prove that u is integral over R, hence in R by hypothesis. This will show that z is in R_1, a contradiction.

Let φ be a place finite on R. If φ is finite on y_1 then it is finite on $h(y_1)$ and it is finite on z which is integral over R_1 by assumption. Hence it is finite on u. If on the other hand φ is infinity on y_1, we divide the numerator and denominator of u by y_1^m where $m =$ deg G. Using the fact that all coefficients of g and G are finite, that deg $g <$ deg G, and that G has leading coefficient 1, we see that the fraction becomes 0 under the place and is therefore finite. This proves what we wanted, and concludes the proof of the proposition.

Note that in this proposition, we have not needed the main hypothesis of ZMT that Q is an isolated point in the image $T(P)$. Still working under the weaker hypotheses of that proposition, it is conceivable that y_1 has a specialization t over $(x) \to P$ which is transcendental over $k(P)$. This would give a subvariety of U_1

(equivalent to a straight line) containing P_1, and corresponding to P on U_1. Since we have proved that T_2 is biholomorphic at P_1, it is biholomorphic on this subvariety, and the image of this straight line on V would be a subvariety of V properly containing Q.

Using now the essential hypothesis of ZMT, that Q is a component in $T(P)$, we can conclude that this cannot happen, i.e. y_1 cannot have such a transcendental specialization over $(x) \to P$. Hence y_1 has only a finite number of specializations over $(x) \to P$, by Proposition 3 of Chapter I, § 4, which must all be algebraic over $k(P)$. It is also conceivable that there is a place mapping (x) on P, and y_1 on infinity. The following proposition now shows that this cannot happen, and in fact that y_1 lies in the local ring of P in $k(x)$. This will therefore conclude the proof of ZMT.

PROPOSITION 4. *Let* $(x) \to (x')$ *be a specialization over k, and assume that the local ring of (x') in $k(x)$ is integrally closed. Let y be an element of $k(x)$ and assume that the number of specializations $y \to y'$ over $(x) \to (x')$ is finite. Then either y or $1/y$ is in the local ring. Hence if y has at least one specialization over $(x) \to (x')$, y must be in the local ring.*

Proof: Let $y_1', \ldots, y_m'$ be the finite number of quantities which may occur as specializations of y over $(x) \to (x')$. There exists a polynomial $f(Y)$ in $k[Y]$ such that $f(y_i') \neq 0$ for any i. Let $w = 1/f(y)$. Then every place inducing $(x) \to (x')$ is finite on w. Indeed, under such a place, if y is infinite, then the denominator of w goes to infinity. If y remains finite, then the denominator of w is not 0, by construction, and so w remains finite. The local ring being integrally closed by assumption, we conclude by Proposition 5 of Chapter I, § 5 that w lies in this local ring. Hence its value is uniquely determined and it is equal to 0 if and only if some place extending $(x) \to (x')$ maps y on infinity.

If every place extending $(x) \to (x')$ is finite on y, then we are through, using the proposition already quoted above. Otherwise, one of these places maps y on infinity, and hence w to 0. Since the value $\varphi(w)$ is unique, this implies that every place extending

$(x) \to (x')$ maps y to infinity, and hence $1/y$ to 0. From this we conclude that $1/y$ is integral over the local ring, and hence in this local ring. This proves our proposition, and finishes the proof of Zariski's Main Theorem.

3. Normalization of an abstract variety

We shall see here that the normalization process extends without difficulty from affine to abstract varieties, simply by normalizing each piece separately, and seeing that they are glued together upstairs in a desirable fashion.

Let $T : U \to V$ be a birational correspondence defined over k between two affine varieties U and V. Let (x) be a generic point of U over k, $T(x) = (y)$, so that $k(x) = k(y) = K$. Let L be a finite algebraic extension of K, regular over k. Let U^* and V^* be the normalizations of U and V in L. Then we have rational maps $f : U^* \to U$ and $g : V^* \to V$, and we can write $L = k(x^*) = k(y^*)$, where (x^*) and (y^*) are generic points of U^* and V^* respectively. There is a birational correspondence $T^* : U^* \to V^*$ such that $T^*(x^*) = (y^*)$. Furthermore $f(x^*) = (x)$, and $g(y^*) = (y)$.

Assume now that T is biholomorphic at a point P of U, and let $T(P) = Q$. We contend that if P^* is any point of U^* lying above P, i.e. such that $f(P^*) = P$, then T^* is biholomorphic at P^*. Indeed, $k[y^*]$ is integral over $k[y]$, and hence integral over the local ring of P in $k(x)$ by assumption. Since the local ring of P^* in $k(x^*)$ is integrally closed by construction, it follows that $k[y^*]$ is contained in this local ring, and hence that T^* is defined at P^*. Let Q^* be the point $T^*(P^*)$. If we apply the preceding argument in reverse, we see that $k[x^*]$ is contained in the local ring of Q^* in $k[y^*]$. From this we conclude that T^* is biholomorphic at P^* and Q^*.

The above remark allows us to define the normalization of an abstract variety $V = [V_\alpha, F_\alpha, T_\alpha^\beta]$ as follows. Let P be a generic point of V over k, and let L be a finite extension of $k(P)$, regular over k. Let V_α^* be the normalization of V_α in L, and let $T_\alpha^{*\beta}$ be the birational correspondence between V_β^* and V_α^* as described in the preceding paragraphs. Let F_α^* be the inverse image of F_α on V^* under the everywhere defined rational map $f^* : V^* \to V$. Then we know that F_α^* is a k-closed set (by Proposition 9 of Chapter IV, § 3). From the above remarks, it follows immediately that $V^* = [V_\alpha^*, F_\alpha^*, T^{*\beta}]$ is an abstract variety, which is called the *k-normalization of V in L*.

In order to characterize V^* up to an isomorphism we recall a definition. Given a rational map $f : U \to V$ from an abstract variety U generically onto an abstract variety V, and defined over k, let P be a generic point of U over k, and let $f(P) = Q$. We say that U is *complete* over a point Q' of V if for every place φ of $k(P)$ which maps Q on Q', some representative P_λ of P is finite under the place and $\varphi(P_\lambda)$ does not lie in the k-closed set G_λ of U_λ. This implies that there is always a point of U lying above the given point of V.

THEOREM 3. *Let V be an abstract variety defined over a field k. Let P be a generic point of V over k, and let L be a finite extension of $k(P)$, regular over k. Then there exists an abstract variety V^* which is defined over k, and is a model of L, and a rational map $f : V^* \to V$, having the following properties:*

1. *V^* is k-normal;*

2. *f is everywhere defined, and there is only a finite number of points on V^* lying above a given point of V;*

3. *V^* is complete over every point of V, so that in particular, there is at least one point on V^* lying above a point of V.*

Any two abstract varieties which are models of L and satisfy these three conditions are k-isomorphic.

Proof: The existence has already been proved. As to uniqueness, let W be an abstract model of L satisfying the three conditions. Then there is a birational correspondence between W and V^*, and to any point on W corresponds at least one point, and at most a finite number of points on V^*. By ZMT, this implies that the map from W to V^* is everywhere defined. The converse is proved in the same manner.

The following proposition shows how the set of local rings of points of V^* in L is completely determined by the set of local rings of points of V in K (cf. the corollaries to Proposition 4 of Chapter I, § 5).

PROPOSITION 5. *Let R be a ring with quotient field K, and let L be a finite extension of K. Let $\mathfrak{p}$ be a prime ideal of R, and let $\mathfrak{o}$ be the local ring $R_{\mathfrak{p}}$. Let $\mathfrak{m}$ be the maximal ideal of $\mathfrak{o}$. Let S be a subring of L, integral over R (and hence over $\mathfrak{o}$) and let $\mathfrak{O}$ be the integral closure of $\mathfrak{o}$ in L. Let $\mathfrak{M}_i$ (with i in some indexing set) be the maximal ideals of $\mathfrak{O}$ lying above $\mathfrak{m}$, and put $\mathfrak{P}_i = \mathfrak{M}_i \cap S$. Then:*

1. *the prime ideals $\mathfrak{P}_i$ are exactly the prime ideals of S lying above $\mathfrak{p}$;*

2. *if S is the integral closure of R in L, then the $\mathfrak{P}_i$ are all distinct, and $S_{\mathfrak{P}_i} = \mathfrak{O}_{\mathfrak{M}_i}$.*

Proof: Since the intersection of $\mathfrak{M}_i$ with $\mathfrak{o}$ is $\mathfrak{m}$, we get immediately $\mathfrak{P}_i = R \cap \mathfrak{p}$, because $\mathfrak{m} \cap R = \mathfrak{p}$. Hence $\mathfrak{P}_i$ is a prime ideal of S lying above $\mathfrak{p}$.

Let $\mathfrak{P}$ be a prime ideal of S which lies above $\mathfrak{p}$. Then there is a homomorphism φ_0 of S which has kernel $\mathfrak{P}$, and whose restriction to R has kernel $\mathfrak{p}$. Extend φ_0 to a place φ of L. Then φ is finite on the local ring $\mathfrak{o}$ and hence on $\mathfrak{O}$. The kernel of the homomorphism of $\mathfrak{O}$ induced by φ is then a prime ideal lying above $\mathfrak{m}$, and must therefore be one of the $\mathfrak{M}_i$. It is then clear that $\mathfrak{P} = \mathfrak{M}_i \cap S$.

Assume finally that S is the integral closure of R in L. The inclusion $S_{\mathfrak{P}_i} \subset \mathfrak{O}_{\mathfrak{M}_i}$ is obvious. Conversely, let x be an element of $\mathfrak{O}_{\mathfrak{M}_i}$. We can write $x = y/z$ where $y, z \in \mathfrak{O}$ and $z \notin \mathfrak{M}_i$. Then y and z are integral over $\mathfrak{o}$. We have an equation

$$y^r + a_{r-1} y^{r-1} + \ldots + a_0 = 0$$

with coefficients $a_j \in \mathfrak{o}$. Each a_j can be written b_j/c_j with $b_j, c_j \in R$, but $c_j \notin \mathfrak{p}$. Hence there exists an element $c \in R$, $c \notin \mathfrak{p}$, (the product of the c_j) such that cy is integral over R. (One sees this by multiplying the above equation by c^r to get $(cy)^r + \ldots + c^r a_0 = 0$.) Hence cy is in S. Similarly, there exists an element $d \in R$, $d \notin \mathfrak{p}$, such that dz is in S. Obviously, dz is not in $\mathfrak{P}_i$. We can write

$$x = \frac{(cy)}{(dz)} \frac{d}{c}$$

and this shows that x lies in $S_{\mathfrak{P}_i}$. This concludes the proof.

From the above proposition, we see that the local rings in L of the normalization of V in L are obtained in the following fashion: For each local ring $\mathfrak{o}$ of V in K, we take the integral closure $\mathfrak{O}$ in L, and then localize with respect to the maximal ideals of $\mathfrak{O}$. This gives a canonical description of this set of local rings, independent of any models selected. In particular, we now see that the property for a rational map to be finite and complete over a point of the base variety is the same for an affine variety or an abstract variety.

PROPOSITION 6. *Let $f : U \to V$ be a separable rational map of degree n, of an abstract variety U generically onto an abstract variety V. Suppose f, U, and V are defined over a field k. Assume that f is everywhere defined on U, and maps U onto V, and that there is at most a finite number of points on U lying above a given point of V. Suppose that both U and V are k-normal. Let V^* be the k-normalization of V in the function field of U. Then the birational correspondence $T : U \to V^*$ is everywhere defined and $1 - 1$. There are at most n points on U lying above a given point of V, and if there are exactly n points, then T is an isomorphism.*

Proof: By ZMT, one sees immediately that T is biholomorphic between U and its image in V^*. By Theorem 1 of Chapter IV, § 5, there can be at most n points on U lying above a given point of V. If there are exactly n points, T must be an isomorphism.

PROPOSITION 7. *Let T be a birational correspondence between two abstract varieties U and V defined over a field k. Assume that U and V are both k-normal, and that for each point P on U, $T(P)$ is not empty and consists of only a finite number of points, and also that for each point Q of V, $T^{-1}(Q)$ is not empty and also consists of a finite number of points. Then T is an isomorphism.*

4. Normalization of a projective variety

The main purpose of this section is to prove the following result.

THEOREM 4. *Let V be a projective variety, defined over a field k. Let P be a generic point of V over k, and L a finite extension of $k(P)$, regular over k. Then the normalization of V in L has a projective embedding, i.e. can be taken to be a projective variety.*

We shall give the entire construction performed before for abstract varieties in the context of projective varieties. We begin with some general results on projective varieties.

We shall work with homogeneous coordinates. By $(x_0, \ldots, x_n)$ we shall always mean a homogeneous generic point, as in Chapter II, i.e. a point such that $(x) \to (tx)$ is a specialization for any transcendental element t over $k(x) = k(x_0, \ldots, x_n)$. The map $(x) \to (tx)$ induces an isomorphism of $k(x)$ over k. Without loss of generality, we may assume that no x_i is 0. For any i, we have $k(x_0/x_i, \ldots, x_n/x_i) = k(x_0/x_j, \ldots, x_n/x_j)$, and we denote this field by $k((x))$.

Geometrically speaking, the points $(x/x_0), \ldots, (x/x_n)$ are ordinary generic points of affine varieties covering the projective variety V whose homogeneous generic point is (x). The function field of V is then $k((x))$.

An element $f(x)$ in $k(x)$ is said to be *homogeneous of degree d* or simply of degree d, if $f(tx) = t^d f(x)$ for t transcendental over $k(x)$. One sees immediately that such an element can be written $f(x) = g(x)/h(x)$ where $g(x)$, $h(x)$ lie in $k[x]$, and $g(X)$, $h(X)$ are forms of degrees a, b respectively such that $a - b = d$. Elements of degree 0 are exactly those of the field $k((x))$. If $f(x) \in k[x]$ is homogeneous of degree d, then one verifies immediately that $f(X)$ is a form of degree d.

The projective variety is *k-normal* if the local rings of its points in $k((x))$ are integrally closed. It is possible to introduce another concept, that of *projective normality*, which should not be confused with the other one. One says that a projective variety is *projectively normal* if the ring $k[x]$ is integrally closed, where (x) is a homogeneous generic point. This property depends strongly on the projective embedding, whereas the property we have been working with before depended only on the local rings of points. Projective normality will be discussed in detail in the

next section. Here we shall only make a few remarks concerning
the relations between the two concepts.

PROPOSITION 8. *Let* (x) *be a homogeneous generic point, and
assume that* $k[x]$ *is integrally closed. Let* $z = \sum c_i x_i$ $c_i \in k$, $z \neq 0$.
Then the ring $k[x/z]$ *is integrally closed.*

Proof: Let $y \in k(x/z)$ be integral over $k[x/z]$, so we can write

$$y^m + g_{m-1}(x/z)y^{m-1} + \ldots + g_0(x/z) = 0$$

and each $g_j \in k[x/z]$. Clear denominators from z. Then we get

$$z^s y^m + h_{m-1}(x)y^{m-1} + \ldots + h_0(x) = 0$$

with $h_j \in k[x]$. After multiplying the above equation by $(z^s)^{m-1}$
we see that $z^s y$ is integral over $k[x]$, hence lies in $k[x]$ by assumption. Write $z^s y = f(x)$. Under the map $(x) \to (tx)$, we know that
y is invariant, and $z \to tz$, so $z^s \to t^s z^s$. Hence $f(tx) = t^s f(x)$. This
implies that f is a form of degree s, and therefore that
$y = z^{-s} f(x) = f(x/z)$ lies in $k[x/z]$, as was to be shown.

Geometrically, the proposition states that whenever the homogeneous coordinate ring is integrally closed, then all the affine
models obtained by dehomogenizing at a hyperplane are k-normal.

The converse of Proposition 8 does not hold. For instance, all
affine models of the projective variety having homogeneous
generic point (t, tu, tu^3, tu^4) have the property that their coordinate rings are integrally closed, but the element tu^2 is integral
over the ring $k[t, tu, tu^3, tu^4]$ because its square lies in that ring,
and yet it does not itself lie in this ring.

The ring $R = k[x]$ generated by the homogeneous point (x)
may be viewed as the ordinary ring of an affine variety, called
the *cone* associated with V. The ideal $Rx_0 + \ldots + Rx_n = \mathfrak{m}$
is a maximal ideal of R (because the residue class ring is obviously
k itself) and corresponds to the point $(0, \ldots, 0)$ on the cone.
If $\mathfrak{p}$ is a prime ideal of R, $\mathfrak{p} \neq \mathfrak{m}$, then $\mathfrak{p}$ is the kernel of a specialization $(x) \to (x')$, with say $x_0' \neq 0$. Put $(y) = (x/x_0)$. Then $(x) \to (x')$
induces the specialization $(y) \to (y') = (x'/x_0')$. Conversely, any
specialization $(y) \to (y')$ can be extended to a specialization
$x_0 \to x_0'$, with $x_0' \neq 0$, and hence $(x) \to (x')$. Thus we see that a

point (x') on the cone with $x_0' \neq 0$ corresponds to a point (y', x_0') one the product variety $V_0 \times D$, where (y) is a generic point of V_0 and x_0 is a generic point of D. Note that x_0 is transcendental over $k(y)$, and we have $k(x) = k(y, x_0)$. Furthermore the local ring of (y', x_0') in $k(y, x_0)$ is the same as the local ring of (x') in $k(x)$.

PROPOSITION 9. *If the projective variety V is k-normal at every point then the cone is k-normal at every point except possibly at the origin.*

Proof: Say we are dealing with a point (x') on the cone with $x_0' \neq 0$. Then $k[y]$ is integrally closed by Proposition 8. The following lemma will show that $k[y, x_0]$ is integrally closed. By Proposition 1 of § 1, we conclude that the local ring is integrally closed.

LEMMA. *Let R be a ring with quotient field K, and let t be transcendental over K. If $\sum a_i t^i \epsilon K[t]$ is integral over $R[t]$, then all a_i are integral over R. If R is integrally closed, then $R[t]$ is integrally closed.*

Proof: Suppose some a_i is not integral over R. There is a place φ of K finite on R and infinity on some a_i. Say a_j has the biggest infinity under the place. We write

$$a_0 + a_1 t + \ldots + a_m t^m = a_j((a_0/a_j) + \ldots + t_j + \ldots + (a_m/a_j)t^m)$$

Since t is transcendental over K, we can extend φ to $K(t)$ and map t on a variable u over $\varphi(K)$. Then using the above representation for $\sum a_i t^i$ we see that all coefficients a_i/a_j are finite under the place, and that $\sum a_i t^i$ maps on infinity under the extended place. However, since this place is finite on $R[t]$, we have a contradiction.

This proves the first part of our lemma. To prove the second part, we note the fact that $K[t]$ is integrally closed (it is an ordinary polynomial ring over a field), and apply the first part.

In order to construct a normalization of the projective variety V, we need to discuss other embeddings of V into projective space.

Let $R = k[x]$ be, as before, a homogeneous coordinate ring for V. Given an integer $d > 0$, we let $\mu_0(x), \ldots, \mu_N(x)$ range over all monomials of degree d in (x). Each $\mu_j(x)$ can be written

$$\mu_j(x) = x_0^{m_0} \ldots x_n^{m_n} \text{ with } \sum m_i = d.$$

We can view $(\mu_0, \ldots, \mu_N)$ as the homogeneous generic point of a projective variety U. We shall now see that U is isomorphic to V.

Let V_0 be the complement of the hyperplane $x_0 = 0$ on V, so V_0 is an affine representative of V having generic point $(1, x_1/x_0, \ldots, x_n/x_0)$ over k. Now dehomogenize U at x_0^d (which is one of the μ_j). We get an affine representative U_0 of U, whose generic point is $(\mu_0/x_0^d, \ldots, \mu_N/x_0^d)$. We contend that $k[x/x_0] = k[\mu(x)/x_0^d]$. The inclusion $\mu_j(x)/x_0^d \in k[x/x_0]$ is trivial. Conversely, $x_1 x_0^{d-1}$ is among the $\mu_j(x)$, and this shows that x_i/x_0 is in $k[\mu(x)/x_0^d]$. From this we see that the affine representative V_0 is essentially the same as the affine representative U_0 of U. It suffices now to remark the obvious fact that any point on U has a representative on some $U_i (i = 0, \ldots, n)$ because not all x_i^d can be 0 simultaneously, otherwise all $\mu_j(x) = 0$.

Phrased another way, the above discussion shows that U gives another projective embedding for V, called the *embedding derived from the hypersurface sections of degree d*. The terminology comes from the fact that each homogeneous element $f(x) \in k[x]$ of degree d defines a hypersurface in projective n-space.

We conclude this section with a proof of Theorem 4. The proof is carried out *a priori*, and makes no use of § 3.

We have $K = k((x))$. Let $(\xi) = (x/x_0)$ and let $S = k[\xi]$. Let J be the integral closure of S in the finite extension L of K. Let $y_1, \ldots, y_r$ be a basis for J over S. Writing down an integral equation for each y_j over $k[\xi]$ and clearing denominators it is immediately verified that for some integer m, we have $x_0^m y_j$ integral over $k[x]$ for $j = 1, \ldots, r$. Repeat this procedure using (x/x_1) and a basis $w_1, \ldots, w_s$ for its integral closure in L. Then $x_1^m w_1, \ldots, x_1^m w_s$ are integral over $k[x]$. (The same m may be used by taking the larger of the two.) Repeat this procedure for each ring $k[x/x_i]$, and let m be an integer which works uniformly for all of them. Let $\mu_0, \ldots, \mu_N$ range over all monomials of degree m in (x). Then by the previous discussion, (μ) is the homogeneous generic point of a projective variety W which is biholomorphic to V.

Consider the point

$$(z) = (\mu_0, \ldots, \mu_N, x_0^m y_1, \ldots, x_0^m y_r, x_1^m w_1, \ldots, x_1^m w_s, \ldots).$$

It is immediately verified that (z) is homogeneous, i.e. that $(z) \to (tz)$ is a specialization for t transcendental over $k(z)$. We can therefore take this point as homogeneous generic point of a projective variety U, whose function field $k((z))$ is obviously equal to L.

We contend that U is normal, and has the required properties of a normalization. (Note that U is not necessarily projectively normal. The question of finding a projectively normal normalization will be dealt with in the next section.) We shall prove our contention in several steps.

1. The ring $k[z]$ is integral over $k[\mu]$ and for any μ_α, $k[z/\mu_\alpha]$ is integral over $k[\mu/\mu_\alpha]$.

Proof: By construction, (z) is integral over $k[x]$, which is integral over $k[\mu]$ (powers of the x's are in $k[\mu]$). This proves the first statement. As to the second, let φ be a place of K finite on $k[\mu/\mu_\alpha]$. We must show φ is finite on (z/μ_α). Suppose φ is infinity on say $x_0^m y_1/\mu_\alpha$. We note that μ_α is transcendental over L, and we can therefore extend φ to $L(\mu_\alpha)$ in such a way that $\varphi(\mu_\alpha) \neq 0, \infty$. We get thereby a place finite on $k[z]$ and infinity on $x_0^m y_1/\mu_\alpha$, hence infinity on $x_0^m y_1$, a contradiction.

2. We recall that for each i, x_i^m is among the μ_α, and we contend that every point on U has a representative on some affine variety with generic point (z/x_i^m).

Proof: Let φ be a place of the function field L of U. Then φ induces a place of $k((x)) = K \subset L$, and hence induces a point on an affine representative of V, say $\varphi(x_i/x_0) \neq \infty$. We know that $k[x/x_0] = k[\mu/x_0^m]$ (trivial), and hence φ is finite on $k[\mu/x_0^m]$. Since x_0^m is among the μ_α, it follows by the first step that φ is finite on $k[z/\mu_\alpha]$ and therefore induces a point on the affine representative whose coordinate ring is $k[z/x_0^m]$, as desired.

From Step 2, we see that to study the relationship between

points of V and points of U, we need only study an affine model of K and its normalization in L. What we have done in our construction of the normalization of V is to show that the piecewise normalizations of affine representatives can be done in such a way that we end up with a projective variety. Furthermore, U is k-normal because each one of the affine representatives in Step 2 is k-normal by construction. This proves the theorem.

Remark that the uniqueness of the normalization can be proved without using ZMT if we deal throughout with projective varieties. We sketch this proof. One first reduces it to the case where k is infinite. Then one knows that if W is another normalization of V which is projective, then to each point P on U corresponds only a finite number of points on W (because we can go from U to W after passing through V). Since k is infinite, we can assume that this finite set of points has a representative on one affine model obtained from the projective variety U by dehomogenizing at a sufficiently general hyperplane, as in Proposition 8. One can then see that the coordinates of these points are integral over the local ring of P, and thus are in this local ring by hypothesis. This shows that the correspondence from U to W is everywhere defined, and hence biholomorphic by symmetry.

For abstract varieties, there was the added difficulty that we did not know how to represent a finite number of points on one affine representative of the abstract variety.

5. Projective normality

As mentioned before, a projective variety with homogeneous generic point (x) over k is called *projectively normal* (with reference to k) if its homogeneous coordinate ring $k[x]$ is integrally closed. This is a stronger condition than normality. We shall study this property, and prove that the normalization of a variety can always be selected projectively normal. We shall give the algebraic background necessary for some projective considerations related to linear systems, and the geometric content of the purely algebraic propositions proved here will appear only in the chapter dealing with linear systems. Hence the reader will do well to postpone reading this section until he reaches the end of Chapter VI.

PROPOSITION 10. *Let $R = k[x]$ be the homogeneous coordinate ring of the projective variety V as before, and let I be its integral closure in $k(x)$. If $f \in I$, then f can be written $f = \sum f_i$, where each $f_i \in k(x)$ is homogeneous of degree $i \geqq 0$, and where also $f_i \in I$. (Some f_i may be 0, of course.)*

Proof: By hypothesis, $f(x) \in k(x)$ satisfies

$$f(x)^m + a_{m-1}(x)f(x)^{m-1} + \ldots + a_0(x) = 0$$

with $a_i(x) \in k[x]$. Since $(x) \to (tx)$ induces an isomorphism of $k(x)$, we have

$$f(tx)^m + a_{m-1}(tx)f(tx)^{m-1} + \ldots + a_0(tx) = 0,$$

and therefore $f(tx)$ is integral over $k[x, t]$ and *a fortiori* over $k(x)[t]$ which is integrally closed. Hence $f(tx) \in k(x)[t]$, and we can write

$$f(tx) = \sum t^i f_i(x) \qquad i \geqq 0,$$

with $f_i(x) \in k(x)$. It must be shown that f_i is homogeneous of degree i. Let u be a new variable. Then tu is transcendental over $k(x)$, and by the above arguments,

$$f(tux) = \sum (tu)^i f_i(x) = \sum t^i u^i f_i(x).$$

We also have

$$f(tux) = \sum t^i f_i(ux).$$

This shows that $f_i(ux) = u^i f_i(x)$, hence that f_i is homogeneous of degree i, as contended. Finally, we shall prove that f_i is integral over R. Indeed, $f(tx)$ is integral over $k[tx]$ and hence over $k[x, t] = R[t]$. That $f_i(x)$ is integral over R now follows from the lemma of Proposition 9 of § 4.

By R_m we shall denote the elements of R which are homogeneous of degree m. Similarly for I_m. We note that R_m and I_m are vector spaces over k, and that R (resp. I) is the direct sum of all spaces R_m (resp. I_m) for $m = 0, 1, \ldots$. This is obvious for R, and is true for I because of Proposition 10.

One sees immediately that $I_0 = R_0 = k$ because k is assumed algebraically closed in $k(x)$. In other words, the elements of $k(x)$ of degree 0 which are integral over R are precisely those of k. Note also that $R_m R_n = R_{mn}$. A similar statement does not necessarily hold for I.

We call m *well behaved* $(m \geq 1)$ if $I_m^\mu = I_{\mu m}$ for all integers $\mu \geq 1$. (By definition, I_m^μ consists of homogeneous polynomials of degree μ in elements of I_m with coefficients in k.) If $R = I$, then all m are well behaved. The following proposition insures the existence of well-behaved integers by showing that all sufficiently large m are well behaved.

PROPOSITION 11. *It is possible to write $I = Rz_1 + \ldots + Rz_s$, where each z_i is homogeneous of degree d_i. Let $m \geq \max d_i$. Then m is well behaved.*

Proof: First, by using Theorem 2 of § 1, we can write I as a finite R module. The elements z_i can always be selected homogeneous, since from given module generators we can derive another set of generators consisting of homogeneous elements (using Proposition 10).

We have trivially for $m \geq \max d_i$

$$I_m = R_{m-d_i} z_1 + \ldots + R_{m-d_s} z_s.$$

Since $R_{m+j-d_i} = R_j R_{m-d_i}$ we get $I_{m+j} = I_m R_j$, and hence if we write $jm = m + (j-1)m$ we obtain

$$I_{jm} = I_m R_{(j-1)m} = I_m R_m^{j-1} \subset I_m I_m^{j-1} = I_m^j \subset I_{jm},$$

thereby proving the proposition.

Note in addition that having expressed I as a finite R module with a homogeneous set of generators, we see that each I_m is a finite-dimensional vector space over k. That R_m is finite-dimensional is obvious.

We shall investigate more closely the ring $k[I_m]$, generated over k by elements of I_m. If $z_0, \ldots, z_s$ is a linear basis for I_m over k, then $k[I_m] = k[z]$. Geometrically, we can view (z) as the homogeneous generic point of a projective variety V_m. If $(w) = (w_0, \ldots, w_s)$ is another linear basis for I_m over k, then the projective variety whose homogeneous generic point is (w) obviously arises from V_m by a projective transformation, so the variety V_m is essentially uniquely determined. The following result shows that V_m is birationally equivalent to V, and that for well-behaved m, the homogeneous coordinate ring $k[z]$ of V_m is integrally closed.

PROPOSITION 12. *The notations being as above, we have* $k((z)) = k((x))$. *If m is well behaved, then $k[I_m]$ is integrally closed.*

Proof: First, z_i/z_0 is homogeneous of degree 0, and hence lies in $k((x))$, so $k((z))$ is contained in $k((x))$. Conversely, since $x_0^m \in I_m$ we can assume without loss of generality that $x_0^m = z_0$. Since $x_0^{m-1} x_i \in I_m$ we can write

$$x_0^{m-1} x_i = \sum_j c_{ij} z_j, \qquad c_{ij} \in k,$$

and hence

$$\frac{x_i}{x_0} = \sum_j c_{ij} \frac{z_j}{z_0}.$$

This shows $k((x)) \subset k((z))$, and hence V_m is birationally equivalent to V.

Assume that m is well behaved. We prove that $k[z]$ is integrally closed. Let $S = k[z] = k[I_m]$. Let J be the integral closure of S in $k(z)$. Note that $J \subset I$ (any element integral over S is *a fortiori* integral over I). We must show $J \subset S$. Let $y \in J$. Apply Proposition 10 to the ring S and to its integral closure J. Then $y = \sum y_i$, where each $y_i \in J$ is homogeneous of degree i in (z). One sees immediately that $k(z)$ is contained in $k(x)$, and that an element w of $k(z)$ which is homogeneous of degree i in (z) is homogeneous of degree mi in (x). Hence we can write $y = \sum_i w_{mi}$ where $w_{mi} \in J \subset I$, and w_{mi} is homogeneous of degree mi in (x). Since m is well behaved, $I_m^i = I_{mi}$. Hence $w_{mi} \in I_m^i \subset k[I_m]$. This proves $y \in k[I_m]$ as desired.

It is easily verified that the birational map $F : V_m \to V$ is everywhere defined on V_m and to every point of V corresponds only a finite number of points of V_m. This gives another method for constructing a normalization of V in its quotient field, which in addition yields a projectively normal variety V_m, and not just a normal one.   In other words, we have proved the following theorem.

THEOREM 5.   *The normalization of a projective variety has a projective embedding in which it is projectively normal.*

We conclude this section with more properties of the ring I. We shall assume from now on that dim $V \geqq 1$, and hence that the dimension of the cone is $\geqq 2$. (Of course, all the preceding results are trivial if dim $V = 0$.)

In the homogeneous coordinate ring $R = k[x]$ we let $\mathfrak{m} = Rx_0 + \ldots + Rx_n$ be the maximal ideal belonging to the origin $(0, \ldots, 0)$. Let $\mathfrak{p}$ denote a prime ideal of R not equal to $\mathfrak{m}$, and such that $R/\mathfrak{p}$ is algebraic over k. We have a local ring $R_\mathfrak{p}$ for each $\mathfrak{p}$ and we define

$$S = \bigcap_{\mathfrak{p} \neq \mathfrak{m}} R_\mathfrak{p}.$$

LEMMA.   *We have* $R \subset S \subset I$.

Proof: The inclusion $R \subset S$ is trivial.   Let $f \in S$.   Let φ be a place of $k(x)$ finite on R.   We shall prove that φ is finite on f. We may assume φ is k-valued.   If $\varphi(x) \neq (0)$, then the kernel of the restriction of φ to R is a prime ideal $\mathfrak{p} \neq \mathfrak{m}$ of R, and since $f \in R_\mathfrak{p}$ we have $\varphi(f) = \infty$.   If $\varphi(x) = (0)$, we would then have a specialization $(x, 1/f) \to (0, 0)$.   This would imply that there exist infinitely many points (x') such that $(x, 1/f) \to (x', 0)$.   Namely, let $w = 1/f$.   View (x, w) as a generic point of an affine variety. Intersect this affine variety with the hyperplane $w = 0$.   By the dimension theorem, all components are of dimension at least equal to $(\dim V + 1) - 1$ (the dimension of the cone is equal to dim $V + 1$), and this is at least 1 by hypothesis.   This implies that there exists a point $(\bar{x})$ of dimension at least 1 over k, such that $(x, 1/f) \to (\bar{x}, 0)$.   We can now take infinitely many algebraic specializations (x') of $(\bar{x})$.   This contradicts the fact that f is in the local ring $R_\mathfrak{p}$ for $\mathfrak{p} \neq \mathfrak{m}$.

PROPOSITION 13.   *For   some   integer   e, $x_i^e S \subset R$ for   each*

$i = 0, \ldots, n$. *In other words*, $\mathfrak{m}^e S \subset R$. *Conversely, if $z \in I$ and $\mathfrak{m}^e z \subset R$ for some e, then $z \in S$.*

Proof: Since $R \subset S \subset I$, we know that S is a finite module over R. Hence to prove our first statement it suffices to show that for each one of a finite number of elements $z \in S$, we have $x_i^e z \in R$ for some e. Say $i = 0$. Given $\mathfrak{p} \neq \mathfrak{m}$, we can write $z = f(x)/g(x)$, and $g(x) \notin \mathfrak{p}$. The ideal generated by the denominators $g(\bar{x})$ in R has therefore only one zero on the cone, and that is the origin $(0, \ldots, 0)$, with corresponding ideal $\mathfrak{m}$. Hence by the Hilbert Nullstellensatz some power $\mathfrak{m}^e$ is contained in this ideal. In particular, we can write

$$x_0^e = \sum h_j(x) g_j(x)$$

with $h_j(x)$, $g_j(x) \in R$, and the $g_j(x)$ occurring as denominators for expressions of z. We observe that $g_j(x)z = f_j(x)$, multiply both sides by h_j, sum over j, and thereby prove that $x_0^e z$ lies in R, as desired.

Conversely, let (x') be a point of the cone with say $x_0' \neq 0$. Its associated ideal $\mathfrak{p}$ is not equal to $\mathfrak{m}$, and by hypothesis $x_0^e z \in R$, and we can write $z = f(x)/x_0^e$ where $f(x) \in R$. This shows that z lies in $R_{\mathfrak{p}}$. This being true for each $\mathfrak{p}$, it follows that z lies in S. This proves the proposition.

PROPOSITION 14.   *The ring S is homogeneous, i.e. if $f \in S$ and $f = \sum f_i$ with $f_i \in I$ homogeneous of degree i (according to Proposition 10) then f_i actually lies in S.*

Proof: According to Proposition 13, $x_0^e f = \sum x_0^e f_i$ lies in R. Each term $x_0^e f_i$ is homogeneous of degree $e + i$. We know that R is homogeneous, and the homogeneous components of an element of R are unique. This proves that each $x_0^e f_i$ lies in R. Similarly, $x_1^e f_i, \ldots$ are in R. Using Proposition 13 again in the reverse direction we conclude that f_i lies in S, as was to be shown.

PROPOSITION 15.   *If as usual S_m denotes the homogeneous elements of S of degree m, then we have $S_m = R_m$ for all sufficiently large m.*

Proof: Since S is homogeneous by Proposition 14, we can write $S = \sum R w_i$, where w_i has degree d_i. Let $d = \max d_i$, and assume

$m \geqq d$. Then clearly

$$S_m = \sum R_{m-d_i} w_i \,.$$

By Proposition 13, $\mathfrak{m}^e S \subset R$, and in addition we obviously have
$R_e \subset \mathfrak{m}^e$. Hence $R_e S \subset R$. Since $R_a R_b = R_{a+b}$ for integers a,
$b > 0$ we get for $m \geqq d + e$

$$S_m = \sum R_{m-d_i-e} R_i w_i.$$

By the above, $R_e w_i \subset R$, and in fact $\subset R_{e+d_i}$. Hence $S_m \subset R_m$.
The inclusion $R_m \subset S_m$ is trivial, and this proves our proposition.

THEOREM 6. *V is k-normal if and only if $I = S$. Hence if V is
k-normal, $R_m = I_m$ for all large m.*

Proof: If V is k-normal we have seen in Proposition 9 that
every point of the cone is k-normal, except possibly the origin.
Hence each $R_{\mathfrak{p}}$ is integrally closed. Hence S is integrally closed.
Since I is the integral closure of R, and $R \subset S \subset I$, it follows that
$S = I$. From this and Proposition 15 we get $R_m = I_m$ for all
large m.

Conversely, assume $I = S$. It suffices to prove that the ring
$k[x/x_0]$ is integrally closed. Let $z \,\epsilon\, k((x))$ be integral over this ring.
Then we have seen many times that $x_0^j z$ is integral over $k[x]$ for
some j. Hence $x_0^j z$ lies in S by hypothesis. By Proposition 13,
for some integer e, $x_0^{e+j} z$ lies in R. Since $x_0^{e+j} z$ is homogeneous of
degree $e + j$ and lies in R, it is a form $f(x)$ of degree $e + j$. Hence
$z = f(x)/x_0^{e+j}$ lies in $k[x/x_0]$ as was to be shown.

Theorem 6 is the main result in the long line of propositions
which have just been developed. Suitably interpreted, it will
mean that the system of hypersurface sections of a normal
projective variety V is complete for sufficiently large m. More
precisely, R_m will later be interpreted as the system of hypersurface sections of V of degree m. Under the assumption that V
is normal, I_m determines the complete linear system arising from
R_m. As to the ring S, its homogeneous elements S_m of degree m
give the global cross sections of the sheaf $o(m)$ defined by Serre
in [4]. The proof is very simple from the definitions. For further

comments on these matters, see the literature at the end of the chapter.

6. Constant field extensions

THEOREM 7. *Let R be a ring containing k, and assume R is integrally closed in its quotient field K. Assume also that k is algebraically closed in K. Let L be a separable extension of k, free from K over k. Then the ring $L[R]$ generated over L by the elements of R is integrally closed.*

Proof: One sees easily that it suffices to prove the theorem stepwise, i.e. assume that L is generated by one element transcendental or separably algebraic.

Case 1. $L = k(t)$ is generated by one transcendental element. An element of $K(t)$ can be written

$$\frac{a_n t^n + \ldots + a_0}{t^m + b_{m-1} t^{m-1} + \ldots + b_0} = \frac{f(t)}{g(t)},$$

with $a_i, b_j \in K$. Without loss of generality, we can assume that f and g are relatively prime in $K[t]$. We shall prove that all the b_j are in k, and all the a_i are in R. This will prove Case 1.

If some b_j is not in k, then b_j is transcendental over k. Hence some root of the polynomial $g(t) \in K[t]$ is transcendental over k (the coefficients are polynomial functions of the roots with integer coefficients), but algebraic over K. There is a place φ of $K(t)$ which is identity on K, and maps t on such a root u of g. Since f and g are relatively prime, u is not a root of f. Hence our quotient f/g is mapped on infinity under the place. However, φ is finite on $k(t)[R]$, a contradiction.

Having all $b_j \in k$, we prove all $a_i \in R$. If some $a_i \notin R$, we can find a place φ of K which is finite on R and infinity on some a_i. Say a_s has the biggest infinity, and write

$$f(t) = a_s\left(\frac{a_n t^n}{a_s} + \ldots + \frac{a_0}{a_s}\right).$$

The place φ takes its values in some field E. We can extend φ to a place of $K(t)$ which maps t on an element u transcendental over E.

Then one sees immediately that φ is finite on $k(t)[R]$ but infinity on f/g, a contradiction.

Case 2. L is finite separable algebraic over k. Let $z \in KL$ be integral over $L[R]$ and hence also over $k[R] = R$. (All elements of L are algebraic over k, hence integral over k and *a fortiori* over R.) Then all conjugates of z over K are integral over R. Let $w_1, \ldots, w_m$ be a linear basis for L over k, and let $w'_1, \ldots, w'_m$ be the dual basis relative to the trace.   Write

$$z = a_1 w_1 + \ldots + a_m w_m, \qquad a_i \in k.$$

Then $a_i = Tr(zw'_i)$. Since each conjugate of zw'_i is integral over R, it follows that a_i is integral over R, hence lies in R by hypothesis. This proves $z \in L[R]$, as desired.

COROLLARY.   *Let $k(x)$ be a regular extension of k, so that (x) is the generic point of a variety V over k. If (x') is a point of V which is k-normal, and if L is a separable extension of k which is free from $k(x)$ over k, then (x') is also L-normal.*

Proof: let $\mathfrak{o}$ be the local ring of (x') in $k(x)$. Then $L[\mathfrak{o}]$ is integrally closed by the theorem. The local ring of (x') in $L(x)$ is the local ring of $\mathfrak{o}$ relative to a prime ideal, and is therefore integrally closed by Proposition 1 of § 1.

We shall now give a proof that the product of two normal points is normal.

LEMMA.   *Let K and L be two extensions of a field k, which are linearly disjoint over k. Assume that K and L are the quotient fields of rings R and S respectively, both of them containing k and integrally closed.   Then the ring $k[R, S]$ in KL is integrally closed.*

Proof: Let I be the integral closure of $k[R, S]$ in its quotient field KL. Then by the preceding theorem, I is contained in $L[R] \cap K[S]$. It suffices now to prove that I is actually equal to this intersection. We view K and L as vector spaces over k. Then R is a direct summand of K over k, and S is a direct summand of L over k, that is, we can write $K = R + R'$ and $L = S + S'$, the sum being direct. It is then an immediate consequence of linear algebra that the only elements of $L[R] \cap K[S]$ are those of $k[R, S]$.

THEOREM 8. *Let V and W be two varieties defined over a field k.*
Let (x′) be a point of V which is k-normal, and let (y′) be a point of W
which is k-normal.   Then (x′, y′) is k-normal on V × W.

Proof: Since V and W are varieties, we can select two independent generic points (x) and (y) such that the fields $k(x)$ and $k(y)$
are free, and hence linearly disjoint by Theorem 3 of Chapter III,
§ 1. If we let R and S be the local rings of $(x′)$ in $k(x)$ and $(y′)$
in $k(y)$ respectively, then $k[R, S]$ is integrally closed. The local
ring of $(x′, y′)$ in $k[x, y]$ is the local ring of $k[R, S]$ relative to a
prime ideal, and hence is integrally closed by Proposition 1 of § 1.

Literature

The central result of this chapter, namely ZMT, is itself a special case of the
principle of connectedness, which asserts that under a correspondence, and suitable
hypotheses of normality, the image of a connected algebraic set is also connected.
The reader will find a proof of this theorem in Zariski's Memoir [7]. He can also
consult various papers of Chow, in course of publication at the time this book is
printed, giving a new approach to these questions.

It was also pointed out by Weil [5] that the fundamental lemma of intersection
theory is an immediate consequence of ZMT.

The normalization of a variety in its function field serves to desingularize the
subvarieties of codimension 1, and was introduced by Zariski partly with this
object in mind.

The normalization in a finite extension of the function field has been used by
Zariski in the above quoted memoir, and more recently by Matsusaka in some
questions dealing with abelian varieties [2], as well as by Rosenlicht in the theory
of algebraic groups [3]. It has also been used in the theory of coverings [1].

The section dealing with projective considerations is entirely due to Zariski,
and can serve as an introduction to his paper on the Lemma of Enriques-Severi [8].
It also gives the algebraic background for some of Serre's projective results in [4].

LANG, S.

[1] Unramified class field theory over function fields in several variables,
     Annals of Math., Vol. 64, No. 2, Sept. 1956, pp. 285—325.

MATSUSAKA, T.

[2] Some theorems on abelian varieties, Natural science report, Ochanomizu
     University, Vol. 4, No. 1, 1953, pp. 22—35.

ROSENLICHT, M.

[3] Some basic theorems on algebraic groups, Am. J. of Math., Vol. 78, No. 2,
     April 1956, pp. 401—444.

SERRE, J. P.

[4] Faisceaux algébriques cohérents, Annals of Math., Vol. 61, No. 2, March 1955,
     pp. 197—278.

WEIL, A.
[5] Foundations of algebraic geometry, Am. Math. Soc. Colloquium publications, Vol. XXIX, New York, 1946, p. 248.

ZARISKI, O.
[6] Foundations of a general theory of birational correspondences, Transactions Am. Math. Soc., Vol. 53, 1943, pp. 490—542.
[7] Theory and applications of holomorphic functions on algebraic varieties over arbitrary ground fields, Memoirs of the Am. Math. Soc. New York, 1951.
[8] Complete linear systems on normal varieties, and a generalization of a lemma of Enriques-Severi, Annals of Math., Vol. 55, No. 3, May 1952, pp. 552—592.

CHAPTER VI

Divisors and Linear Systems

By the *codimension* of a subvariety W of a variety V we mean $\dim V - \dim W$.

The subvarieties of codimension 1 on a normal variety are of considerable importance because they can be handled to a certain extend by the methods which apply to curves, i.e. discrete valuations. It is possible to define the order of the zero and pole of a function on such a subvariety to be an integer, and one can then define the divisor of a function in the group of all divisors, which is the free abelian group generated by all the subvarieties of codimension 1. If the variety is complete, then the space of functions which have at most a given finite set of poles (with multiplicities) is finite-dimensional over the constant field, and gives rise to a rational map of V into projective space. Questions of projective embedding are then equivalent to questions concerning linear systems which arise from such finite-dimensional vector spaces.

All of this chapter will take place over a fixed ground field k, algebraically closed, except the last section, where we see *a posteriori* that the theory can be extended to arbitrary ground fields. Chapter III will be used only in this last section.

1. Divisors and divisors of functions

Let V be a variety of dimension r, defined over a field k. We assume first that V is affine, since we shall begin by studying a local theory. We let (x) be a generic point of V over k, and put $K = k(x)$. We call K a function field of V.

THEOREM 1. *Let (x') be a specialization of (x) over k, and assume that (x') is of dimension $r - 1$ over k, and that its local ring in $k(x)$ is integrally closed. Then this local ring is a valuation ring.*

Proof: Let $\mathfrak{o}$ be the local ring. Let $z \in k(x)$, and $z \notin \mathfrak{o}$. We shall prove that $1/z$ is in $\mathfrak{o}$. Since $\mathfrak{o}$ is integrally closed, it follows that $(x, 1/z) \to (x', 0)$ is a specialization, i.e. there is a place sending (x) to (x') and z to infinity. On the other hand, we contend that z has only a finite number of specializations over $(x) \to (x')$. Indeed, if it had an infinite number, then $(x, z) \to (x, t)$ would be a specialization, where t is transcendental over $k(x')$ by Proposition 3 of Chapter I, § 4. But $\dim_k(x', t) = r$ would mean that this specialization yields an isomorphism of $k(x)$, which is impossible. Our theorem is now a consequence of Proposition 4 of Chapter V, § 2. (The reader will note however that the proof of that proposition is self-contained with respect to Chapter I, and no formidable machinery involving ZMT is involved. The proposition was put in Chapter V only because it was not needed before then, but it could have been placed in Chapter I.)

It is convenient to mention here a corollary involving some global considerations.

COROLLARY. *Let $f : V \to W$ be a rational map of an affine variety V into a complete abstract variety W. If V' is a subvariety of codimension 1 which is normal on V, then f is defined at V', i.e. at a generic point of V' over a common field of definition for f, V, W, and V'.*

Proof: Let (x) be a generic point of V over k, and (x') a generic point of V' over k. The local ring $\mathfrak{o}$ of (x') in $k(x)$ is a valuation ring. Let φ be a place belonging to this valuation ring. Let $Q = f(x)$. Let $W = [W_\gamma, G_\gamma, R_\lambda^\gamma]$. The point Q has representatives Q_γ, and for some γ, $\varphi(Q_\gamma) = Q'_\gamma$ is the representative of some point Q' of W, because W is assumed to be complete. If $Q_\gamma = (y_1, \ldots, y_m)$ then by definition of a valuation ring, all $y_j \in \mathfrak{o}$, whence f is defined at (x').

Let now W be a subvariety of V, of dimension $r - 1$, and defined over k. Let (x'), be a generic point of W over k, so $\dim_k(x') = r - 1$. If the local ring of (x') in $k(x)$ is integrally closed, it is a valuation ring. If it is not, then we still have

PROPOSITION 1.   *Given a specialization $(x) \to (x')$ where $\dim_k(x') = r - 1$. Then there exists only a finite number of places of $k(x)$ extending this specialization.*

Proof: One can give two arguments. The first uses a result of Chapter V, namely that the integral closure of $k[x]$ in its quotient field is finitely generated. One sees that the places extending $(x) \to (x')$ are in $1 - 1$ correspondence with the homomorphisms of the integral closure inducing $k[x] \to k[x']$, and these must be finite in number. The other argument is independent of the finiteness theorem of Chapter V. Let $x'_1, \ldots, x'_{r-1}$, say, be algebraically independent over k. Then so are $x_1, \ldots, x_{r-1}$. Let $K_0 = k(x_1, \ldots, x_{r-1})$. Then any place φ inducing $(x) \to (x')$ is an isomorphism on K_0. Without loss of generality, we may assume that it is identity on K_0. Note that $k(x)$ is a finite algebraic extension of $K_0(x_r)$ (if say x_r is transcendental over K_0). We can then apply the corollary to Theorem 4 of Chapter I, § 6.

Assume for the rest of this section that k is algebraically closed. All subvarieties W of V of codimension 1 will also be assumed to be defined over k.

A subvariety W of V as above will be said to be *non-singular* if the local ring in $k(x)$ of a generic point (x') of W over k is a valuation ring. The chapter on simple points later will justify this terminology. By a *divisor* on V we shall mean an element of the free abelian group generated by the non-singular subvarieties of codimension 1, and defined over k. In Section 4, we shall explain how the theory can be generalized to an absolute theory over the universal domain. By a *prime divisor* we shall mean a non-singular subvariety of codimension 1.

If V^* is the normalization of V in its function field, then there may correspond several subvarieties of codimension 1 on V^* to one such subvariety of V. If the reader wishes to make our discussion independent of Chapter V, we might take as definition of *prime divisor on the normalization V^* of V* the valuation rings in $k(x)$ which induce a specialization $(x) \to (x')$ where (x') has dimension $r - 1$ over k. Proposition 1 guarantees that there is only a finite number of them. The free abelian group generated by these prime divisors is then called the group of divisors on the normalization V^* of V.

Let W be a prime divisor of V, and let $\mathfrak{o}$ be its valuation ring in $k(x)$. Let z be a function in $k(x)$. Then either z or z^{-1} lies in $\mathfrak{o}$. If z does not lie in $\mathfrak{o}$, then z^{-1} lies in the maximal ideal $\mathfrak{m}$ of $\mathfrak{o}$, and in that case we shall say that z has a *pole* at W. If z itself lies in $\mathfrak{m}$, we say that z has a *zero* at W. We see immediately that z has a zero at W if and only if there is a specialization $(x, z) \to (x', 0)$ where (x') is a generic point of W over k. If φ is a place of $k(x)$ whose valuation ring is $\mathfrak{o}$, then z has a pole or zero at W according as $\varphi(z) = 0$ or $\varphi(z) = \infty$, whence the terminology.

We have considered above functions in a function field $k(x)$ of V. Of course, we can make our results geometric by considering functions as rational maps of V into S^1. If f is such a function, and $f(x) = z$, then we say f has a *pole* or a *zero* at W according as z has a pole or zero. In view of the natural isomorphisms between the function fields of V obtained by taking different generic points, this definition is clearly independent of the particular field selected. For the rest of this section, we identify frequently a function f in *the* function field, with its value $f(x)$ in *a* function field $k(x)$ of V over k.

If f is a function defined over k, and $z = f(x)$, then the locus of (x, z) over k is called the *graph* of the function, and is denoted by Γ_f. If we intersect Γ_f with $V \times 0$, then all components have dimension $r - 1$, because they are precisely the components of $\Gamma_f \cap (S^n \times 0)$ taken on the product affine space $S^n \times S^1$. One sees immediately that the components which are non-singular are precisely the zeros of the function (more precisely that their projections on the first factor are the zeros of the function). The singular components will correspond to zeros of the function on the normalization of V. In particular, we get

PROPOSITION 2. *A function has only a finite number of zeros and poles.*

By a similar argument we also prove the following result.

PROPOSITION 3. *Assume V is non-singular in codimension* 1, *and let (y) be a point on V whose local ring is integrally closed (i.e. a normal point). If a function f is not defined at (y), then f has a pole W containing (y).*

Proof: Since the local ring of (y) is assumed integrally closed, there is a place of $k(x)$ mapping (x) on (y) and $z = f(x)$ on infinity. Let $t = z^{-1}$, and let Γ be the graph of f^{-1}. Then the intersection $\Gamma \cap (V \times 0)$ is not empty because $(y, 0)$ is in it, and all components are of dimension $r - 1$ by the dimension theorem. One of these components must contain $(y, 0)$, and from the definitions we see that the projection of this component on V must be a pole of our function.

We shall now investigate the value group of the valuation ring associated with a prime divisor. We shall prove that it is canonically isomorphic to the integers, and this will allow us to speak of the order of the zero or pole that a function may have at a prime divisor.

Consider first the case where $(x) = (x_1, \ldots, x_r)$ consists of r independent variables over k. Then we have unique factorization in $k[x]$ and we have already studied the valuation rings associated with the primes of this ring (Chapter I, § 3). Note in addition to the facts mentioned there that if W is a prime divisor and (x') is a generic point of W, then the kernel of the homomorphism $k[x] \to k[x']$ is a prime ideal of $k[x]$, generated by one of the irreducible polynomials $p(x)$ of $k[x]$. Hence if we write an element $z \in k(x)$ in the form $z = p(x)^m R(x)$, where $R(x)$ is a rational function whose numerator and denominator are not divisible by $p(x)$ (we say that $R(x)$ is prime to p) then z has a zero or pole according as $m > 0$ or $m < 0$. We write $m = ord_W z$.

We pass to the general case, with $(x) = (x_1, \ldots, x_n)$. Let (x') be a generic point of a prime divisor W on V. Say $x_1', \ldots, x_{r-1}'$ are algebraically independent over k. Then so are $(x_1, \ldots, x_{r-1})$. Say x_r is independent from $k(x_1, \ldots, x_{r-1})$. Then the place over $(x) \to (x')$ induces a place on the subfield $k(x_1, \ldots, x_r)$ of $k(x)$, and may be viewed as inducing a prime divisor on affine space S^r. We have seen that the value group from the subfield is infinite cyclic. The value group of $k(x)$ must also be infinite cyclic by Proposition 7 of Chapter I, § 6.

This infinite cyclic value group has a canonical generator, arising from the value $v(t)$ of an element t which lies in the valuation ring. Hence it is canonically isomorphic to the integers.

All ideals of the valuation ring $\mathfrak{o}$ are or type $\mathfrak{m}^s$ where s is some positive integer. Indeed, any element of $\mathfrak{o}$ can be written $t^s u$, where u is a unit in $\mathfrak{o}$. So an ideal is generated by a power t^s of an element t whose value $v(t)$ generates the value group, and $\mathfrak{m}$ itself is generated by t.

If t is an element of $\mathfrak{o}$ such that $v(t)$ generates the value group, and if z is an element of $k(x)$ which can be written $z = t^s u$, where s is an integer and u is a unit in $\mathfrak{o}$, then we define $s = ord_W z$. If we have a function f such that $f(x) = z$, then s is the *order of f at W* (or of z at W). This definition is clearly independent of the choice of function field $k(x)$.

The reader should note that a place of $k(x)$ over k whose residue class field is of transcendence degree $r - 1$ over k is not necessarily induced by a prime divisor of V. For example, we let t, u be two independent variables over k. Then $(t, u, t/u) \to (0, 0, v)$ is a specialization for v transcendental over k. Its local ring is a valuation ring inducing a prime divisor of the locus of $(t, u, t/u)$ over k, but inducing a point on the locus of (t, u) over k, namely the point $(0, 0)$.

Let now V be an abstract variety. We recall that this means that V consists of $[V_\alpha, F_\alpha, T_\alpha^\beta]$, where F_α is an algebraic subset of V_α and not equal to V_α, and T_α^β are birational correspondences between V_β and V_α satisfying the following condition: If P'_β and P'_α are points of V_β and V_α corresponding under T_α^β and such that $P'_\alpha \notin F_\alpha$ and $P'_\beta \notin F'_\beta$, then T_α^β is biholomorphic at P'_β and P'_α. If P_α is a generic point of V_α over k, and P_β is a generic point of V_β over k such that $P_\alpha = T_\alpha^\beta(P_\beta)$, then our condition means that the local ring of P'_α in $k(P_\alpha)$ is equal to the local ring of P'_β in $k(P_\beta)$. Of course, the two fields $k(P_\alpha)$ and $k(P_\beta)$ are the same.

If P'_α has dimension $r - 1$ over k and is not in F_α then it is the representative on V_α of a generic point P' of a subvariety W of V. Everything we have done in the preceding discussion depended only on the local ring of P' in $k(P)$, and thus we see that our results apply to an abstract variety. In addition, we shall now prove one more result which holds for complete abstract varieties.

PROPOSITION 4. *Let V be an abstract variety, non-singular in codimension 1, and complete. If a function on V is not a constant, then it has at least a zero and a pole.*

Proof: Let P be a generic point of V, and $z \in k(P)$, $z = f(P)$. By hypothesis, $z \notin k$. We map z on 0, and extend this to a place of $k(P)$ over k. On one of the affine representatives V_α of V, the place will induce a point not in the frontier. If P' is such a point, this means that $(P', 0)$ is contained in the graph Γ of f, and hence that $\Gamma \cap (V \times 0)$ is not empty. We are therefore reduced to a local property of affine varieties, and we can repeat the argument used in Proposition 2 or 3, in order to get a zero for f. To construct a pole, we need only construct a zero for f^{-1}.

2. Existence of functions with given zeros

We shall here prove a result which will be used in Theorems 2 and 6.

PROPOSITION 5. *Let V be an affine variety, and $W_i (i = 1, \ldots, n)$ a finite number of distinct prime divisors of V, all defined over an algebraically closed field k. Let $R = k[x]$ be a coordinate ring for V. Let m_i be integers ≥ 0. Then there exists an element $y \in R$ such that $\mathrm{ord}_{W_i}(y) = m_i$.*

Proof: Let $\mathfrak{p}_i$ be the prime ideals in R belonging to the W_i. There cannot be any inclusion relations among them because there is none among the W_i, by Corollary 1 of Theorem 5, Chapter II, § 2. We can find $y_j \in R$ such that $y_j \notin \mathfrak{p}_1$, but $y_j \in \mathfrak{p}_j$, for $j = 2, \ldots, n$. The product $\prod y_j = y$ is a unit at W_1 (*i.e.* has no zero or pole at W_1) and has a zero at $W_j (j = 2, \ldots, n)$. A high power of this y is a unit at W_1 and has a zero of high order at W_j. Thus for each i, we can find a function which is a unit at W_i and has a zero of high order at W_j for $j \neq i$.

Now let $w_1, \ldots, w_n$ be elements of R such that each w_i has order m_i at W_i. If $m_i = 0$, all we need to take is $w_i \in R$ and $w_i \notin \mathfrak{p}_i$. Otherwise, the ideal $\mathfrak{p}_i$ generates the maximal ideal of the local ring of W_i (which is the valuation ring of W_i) and hence we can find an element in R which has order 1 at W_i. The m_ith power of this element has order m_i at W_i. Let $z_1, \ldots, z_n$ be elements of R such that z_i is a unit at W_i and has a zero of high order at $W_j, j \neq i$. Then $z_1 w_1 + \ldots + z_n w_n$ satisfies our requirements.

3. Linear systems

Let V be an abstract variety of dimension r, defined over the algebraically closed field k. A divisor, i.e. an element of the free abelian group generated by the non-singular subvarieties of codimension 1 (and defined over k for this section and the next) can be written

$$D = \sum n_i W_i,$$

where n_i are integers. If $D \neq 0$, we can write D in a unique manner such that the W_i are distinct, and the $n_i \neq 0$. If all $n_i > 0$, we say that the divisor is *positive* and we write $D > 0$. Given two divisors D_1 and D_2, we write $D_1 > D_2$ if $D_1 - D_2 > 0$.

Given a function f on V we can associate a divisor with it in the following manner. For every non-singular prime divisor W of V we have defined in § 1 the integer $\mathrm{ord}_W f$. We know that a function has only a finite number of zeros and poles. Hence we may associate with f the divisor

$$(f) = \sum n_i W_i,$$

where $n_i = \mathrm{ord}_{W_i} f$, and the sum is taken over all prime divisors at which f has a zero or a pole.

The sum $\sum n_i W_i$ taken only over those prime divisors where f has a zero is called the *divisor of zeros* of f. The sum $- \sum n_i W_i$ taken only over those prime divisors where f has a pole is called the *polar divisor* of f. They are denoted by $(f)_0$ and $(f)_\infty$ respectively, and are both ≥ 0. Similarly if D is a divisor, and one takes only the sum of those terms where a prime divisor occurs with negative multiplicity, then one gets the *polar part* of D.

Two divisors D_1 and D_2 are called *linearly equivalent* if they differ by the divisor of a function, i.e. if there is a function f such that $D_1 = (f) + D_2$.

Given a divisor D, we denote by $\mathscr{L}(D)$ the set of all positive divisors linearly equivalent to D. It may be empty. The set of all functions f such that $(f) \geq - D$ is clearly a vector space over k which is denoted by $L(D)$. (We regard the constant function 0 as having a divisor $> D$ for all D.) The dimension of this space is denoted by $l(D)$. It may be infinite.

If $D > 0$ this space consists of all functions having at most poles of D, counting multiplicities.

If P is a generic point of V over k, then $k(P)$ is isomorphic to the function field of functions f on V defined over k under the mapping $f \to f(P)$. We can obviously associate a divisor to a function $z \, \epsilon \, k(P)$, and the above definitions for functions apply to elements of $k(P)$. By an abuse of notation, we shall sometimes use $L(D)$ to denote the space of functions in $k(P)$ satisfying $(z) \geq - D$.

We shall prove below that if V is complete and non-singular in codimension 1, then $l(D)$ is finite. Here we make some preliminary remarks. If $D = 0$, then by Proposition 4 of § 1 we know that $L(D)$ consists of the constants alone. Furthermore, if $D_1 \geq D_2$, then obviously $L(D_1) \supset L(D_2)$. Hence to prove this result, it will suffice to consider positive divisors. We now come to the theorem.

THEOREM 2. *Let D be a divisor on a complete abstract variety V, non-singular in codimension 1. Then the vector space $L(D)$ of functions f such that $(f) \geq - D$ is finite-dimensional.*

Proof: The proof will be carried out by induction on the dimension r of V.

Assume first that V is a curve (i.e. has dimension 1). The prime divisors are then the points of V. We may clearly assume that $D \geq 0$, and it will also suffice to prove that if the theorem is true for some divisor D, then it is true for $D + Q$, where Q is a point. In fact we shall prove that $l(D + Q) \leq l(D) + 1$. Suppose that there exist two functions z and $w \, \epsilon \, L(D + Q)$ which are linearly independent mod $L(D)$. Let n be the multiplicity of Q in D. Since z and w are not in $L(D)$, we must have $\mathrm{ord}_Q z = \mathrm{ord}_Q w$

$= n + 1$. Hence z/w is a unit in the valuation ring, and, under the place associated with the valuation ring, takes on some constant value $a \in k$. Then $zw^{-1} - a$ has a zero at Q, and $w(zw^{-1} - a)$ $= z - aw$ has a pole of order at most n at Q. But $z - aw$ lies in $L(D + Q)$ and hence in $L(D)$. This contradicts the linear independence of z and w mod $L(D)$.

Let now V have dimension r and let W be a subvariety of codimension 1 on V. Assume that our theorem has been proved for $r - 1$. We shall show how to prove it for V. We have $V = [V_\alpha, F_\alpha, T_\alpha^\beta]$. and for simplicity, we shall assume that all F_α are empty (as one can assume without loss of generality by Proposition 17 of Chapter IV, § 6.)

Let P be a generic point of V over k, and let P_α be corresponding representatives on the affine varieties V_α. Let W be a prime divisor on V, and let P' be a generic point of W. For some α, P' has a representative P'_α. Suppose we have a function $f(P) = z$ in $k(P)$ which has no pole at W. Then z is in the local ring of P'_α in $k(P_\alpha)$, which is the valuation ring of W. Hence z is defined at P'_α, and has a uniquely determined specialization $z \to z' = f(P')$ over $P_\alpha \to P'_\alpha$. It is clear that the function z' in $k(P')$ which we obtain thus is independent of the representative P'_α selected for W, i.e. if P' has a representative P'_β on another V_β, then viewed as a function of P_β, z is in the local ring of P'_β in $k(P_\beta)$ and has the specialization z' over $P_\beta \to P'_\beta$. The function z' in $k(P')$ will be said to have been induced by z on W. The above process of inducing a function on a subvariety W of codimension 1 will be used in the induction, which depends on the following lemma.

LEMMA. *Let V be complete and non-singular in codimension 1. Let D be a divisor on V. Let W be a prime divisor of V, not contained in any component of D. Then there exists a divisor D^* on the normalization W^* of W such that for any function $z \in L(D)$, the induced function z' on W satisfies $z' \in L(D^*)$.*

Proof: Our statement depends on the fact that W is not a pole of z, and hence that z is defined at W and therefore induces a function z' as described above.

Let R_α be a coordinate ring of V_α.

By Proposition 5, there exists a function y_α in R_α such that for any $z \in L(D)$ the function $y_\alpha z$ has no pole having a representative on V_α, and such that y_α is a unit in the valuation ring of W in $k(P)$, i.e. has no zero or pole at W. All we need to find is an element y_α of R_α which has order 0 at W, and has a zero of high order at all the components of D appearing with negative multiplicity in D.

If W has a representative on V_α, then the functions y_α, z, and $y_\alpha z$ are defined at W, and induce functions y'_α, z', and $y'_\alpha z'$ on W. Furthermore $y'_\alpha \neq 0$, and we have $(y_\alpha z)' = y'_\alpha z'$. If W_α is the affine representative of W on V_α, then the function $y'_\alpha z'$ cannot have a pole on the normalization W^*_α of W_α. This is easily proved as follows. Let P'_α be representative of the generic point of W on V_α. If our function $y'_\alpha z'$ had such a pole, we would have a specialization $(P'_\alpha, 1/y'_\alpha z') \to (P''_\alpha, 0)$, where P''_α is the generic point of a subvariety of codimension 1 on W. On the other hand, if P_α is the representative of P on V_α, we have a specialization $(P_\alpha, 1/y_\alpha z) \to (P'_\alpha, 1/y'_\alpha z')$. Composing the two specializations, and using an argument as in Proposition 3, we see that $y_\alpha z$ has a pole on V_α, contrary to assumption.

One sees from the above that the poles of z' on W having a representative on W_α are bounded by the zeros of y'_α. Since W is covered by a finite number of these affine varieties, it follows that z' has bounded poles, as desired. (Strictly speaking, these poles and zeros of z' are of course on the normalization W^* of W, but as we have seen, there is only a finite number corresponding to a given subvariety of codimension 1 on W.)

We return to the proof of the theorem. For $D = 0$, we know that $L(0) = k$ is the constant field, by Proposition 4. It will therefore suffice to prove that if $L(D)$ is finite-dimensional for $D \geq 0$ then for any prime divisor W, $L(D + W)$ is also finite-dimensional.

Let W appear with multiplicity $n \geq 0$ in D, so $D = nW + \ldots$. If there is no function $y \in L(D + W)$ having a pole of order $n + 1$ at W, then $L(D + W) = L(D)$ and we are done. Otherwise, let

$y \in L(D + W)$ be such that $\operatorname{ord}_W y = -(n + 1)$, i.e. y has a pole of order $n + 1$ at W. Consider the vector space $y^{-1}L(D + W)$ consisting of all products $y^{-1}z$ with $z \in L(D + W)$. It is obviously equal to $L(D + W + (y))$. None of the functions in it has a pole at W. Hence all of these functions induce functions on W, and we have a homomorphism $w \to w'$, where $w \in L(D + W + (y))$ and w' is the function induced by w on W. According to the lemma, and by the induction hypothesis, the dimension of the image space under our homomorphism is finite. We must therefore show that the kernel has finite dimension.

Suppose that a function $y^{-1}z$ maps onto 0. This means that $y^{-1}z$ has a zero at W, and hence that z has a pole of order at most n at W. Since $z \in L(D + W)$, we see that in fact $z \in L(D)$. Thus the kernel of our homomorphism is k-isomorphic to $L(D)$, which is assumed to be finite-dimensional. This concludes the proof of the theorem.

For the rest of this section and also the next section we assume that V is complete and non-singular in codimension 1.

We have worked before with functions in a function field $k(P)$ of V over k. We now go back to the more geometric terminology.

Let $\mathscr{L}$ be a non-empty set of linearly equivalent positive divisors. Let X_0 be a divisor in $\mathscr{L}$. For each $X \in \mathscr{L}$, $X - X_0$ is the divisor of some function f, and any two such functions differ by a constant. If $(f) = X - X_0$, then $(f) \geqq - X_0$, so $f \in L(X_0)$. Consider the following condition:

The set of all functions f such that $(f) = X - X_0$ for some $X \in \mathscr{L}$ is a vector space L_0 over k. (By Theorem 2 this space is necessarily finite dimensional.)

If X_1 is another divisor in $\mathscr{L}$, then it is immediately verified that the set of functions f such that $(f) = X - X_1$ for some $X \in \mathscr{L}$ also forms a vector space L_1. In fact, if g is a function such that $(g) = X_0 - X_1$, then $L_1 = gL_0$.

If the set $\mathscr{L}$ of positive divisors satisfies the above condition, then $\mathscr{L}$ is called a *linear system*. For any divisor $D \neq 0$, the set of all positive divisors linearly equivalent to D is a linear system, provided it is not empty. It is denoted by $\mathscr{L}(D)$. It can also be described as the set of all positive divisors in a linear equivalence class of divisors. Such a linear system is called *complete*.

Let L be a given finite-dimensional space of functions. We can associate with L a linear system as follows. Let $f_0, \ldots, f_n$ be a basis for L over k. Each f_i has a divisor

$$(f_i) = X_i - Y_i,$$

where X_i and Y_i are the divisors of zeros and poles of f_i respectively. Any function of L is a linear combination of the f_i. We see immediately that the sup of all Y_i is the sup of all polar divisors of functions $f \epsilon L$. Let $Y = \sup Y_i$. All $f \epsilon L$ satisfy $(f) \geqq - Y$ (but there may be functions g such that $(g) \geqq - Y$ and $g \notin L$). With the vector space L we associate the set of positive divisors consisting of all divisors $(f) + Y$, with $f \epsilon L$. This is clearly a linear system. Conversely, a linear system $\mathscr{L}$ determines a class of spaces gL as discussed before, any one of which is said to be associated with $\mathscr{L}$.

We now give an example of a linear system on a projective variety, namely the linear system of hypersurface sections of a given degree.

Let V be a projective variety, let $R = k[x]$ be a homogeneous coordinate ring, as in Chapter V, § 5, and let R_m be the homogeneous elements of degree m in R. Let $f \epsilon R_m$. Let W be a prime divisor on V. Say W is not contained in the hyperplane $x_0 = 0$. Then $f(x)/x_0^m$ determines a function φ on V, and we shall associate with f the divisor $Z(f)$ containing W with multiplicity $\mathrm{ord}_W\varphi$. (It is clear that $\mathrm{ord}_W\varphi$ actually does not depend on the selected hyperplane used to dehomogenize V, and that there is only a finite number of prime divisors where $\mathrm{ord}_W\varphi$ is not 0.) The following statements now follow trivially from the definitions.

All the divisors $Z(f)$, with $f \epsilon R_m$, are linearly equivalent. In fact if $f, g \epsilon R_m$, then $(f/g) = Z(f) - Z(g)$.

If $f_0 \epsilon R_m$, then the set of all functions f/f_0, $f \epsilon R_m$, constitutes a vector space L over k, and the sup of all polar divisors of functions in L is $Z(f_0)$. Thus the set of divisors $Z(f)$, $f \epsilon R_m$, is a linear system, defined by the above vector space. It is called the linear system of hypersurfaces of degree m.

Let I be the integral closure of R in its quotient field, and let I_m be the homogeneous elements of I of degree m (cf. Proposition 10 of Chapter V, § 5). Then we have the following theorem of Zariski:

THEOREM 3. *If φ is a function on V such that $(\varphi) \geqq - Z(f)$ for some $f \in R_m$, then there exists $g \in I_m$ such that $\varphi = g/f$.*

Theorem 6 of Chapter V, § 5 can now be interpreted as asserting that the linear system of hypersurfaces of degree m on a normal variety is complete for all sufficiently large m.

The proof of Theorem 3 will be left as an exercise to the reader.

4. The rational map associated with a linear system

For this entire section, we assume that V is a complete abstract variety, non-singular in codimension 1, and defined over the algebraically closed field k. All subvarieties of codimension 1 are likewise assumed defined over k.

Let $\mathscr{L}$ be a linear system on V. We have seen that $\mathscr{L}$ determines a class of vector spaces of functions in the following manner. If $X_0 \in \mathscr{L}$, then L_0 consists of all functions f such that $(f) = X - X_0$ for some X in $\mathscr{L}$. If L is any space of functions obtained in this fashion, then $L = gL_0$ for some function g.

Let L be any one of the vector spaces derived in the above manner. Let P be a generic point for V over k. Let $f_0, \ldots, f_n$ be a basis for L over k, and put $f_i(P) = y_i$. Let t be a quantity transcendental over $k(P)$. Then $(ty_0, \ldots, ty_n)$ is the homogeneous generic point in projective n-space of a variety V', and defines a rational map $T : V \to V'$ of V generically onto the projective variety V'. A change of basis for L merely gives a projective transformation of V', and if another module L_1 derived from $\mathscr{L}$ is used instead of L, we see again that we get the same rational map, up to a projective transformation. Hence we may say that the rational map depends only on $\mathscr{L}$. It is said to be *induced*, or *associated with $\mathscr{L}$*.

A positive divisor Y is said to be a *fixed divisor* of the linear system $\mathscr{L}$ if $X \geq Y$ for all $X \in \mathscr{L}$. If we subtract from all divisors of $\mathscr{L}$ a fixed divisor of $\mathscr{L}$, then we get again a linear system (this follows immediately from the definitions). Furthermore, the rational map which we get from the new linear system is clearly the same as that from the original one. Hence to study the rational map induced by a linear system, we may suppose without loss of generality that there is no fixed divisor.

In the next proposition, we give a criterion for the rational map to be defined at a point. A point P of V is said to be a *base point* of the linear system if $P \in \operatorname{supp} (X)$ for all $X \in \mathscr{L}$. Every point in a fixed divisor is a base point, but it may well happen that a linear system has no fixed divisor but has a base point.

PROPOSITION 6. *Let $\mathscr{L}$ be a linear system without fixed divisor, and let T be the rational map of V defined by $\mathscr{L}$. Let P be a point normal on V. Then T is defined at P if and only if P is not a base point of $\mathscr{L}$.*

Proof: Suppose P is not a base point. Let X_0 be an element of $\mathscr{L}$ such that $P \notin \operatorname{supp} (X_0)$. Let L_0 be the space of functions obtained by dehomogenizing $\mathscr{L}$ at X_0, and let $f_0, \ldots, f_n$ be a basis for L_0 as before. We have $(f_i/f_0) = X_i - X_0$ and by Proposition 3 of § 1, we see that f_i/f_0 has no pole through P and is defined at P. The converse is equally trivial.

Let $\mathscr{L}$ again be a linear system without fixed divisor, and let $T : V \to V'$ be the rational map induced by $\mathscr{L}$. We shall see that T maps divisors on hyperplanes in the following manner. Let $X \in \mathscr{L}$. Let $P \in \operatorname{supp} (X)$. We assume that P is not a base point, and we select a divisor $X_0 \in \mathscr{L}$ such that $P \notin \operatorname{supp} (X_0)$. Let L be the vector space derived from $\mathscr{L}$ at X_0. Say $f_0, \ldots, f_n$ is a basis for this space. Then any function in L can be written $f = f_c = \sum c_i f_i$ with $c_i \in k$, and $(f_c) = X_c - X_0$. Say $X = X_c$. Since P is not in $\operatorname{supp} (X_0)$, all functions f_i are defined at P. Since $P \in \operatorname{supp} (X_c)$, it follows that $\sum c_i f_i(P) = 0$, i.e. that P gets mapped into the hyperplane defined by the coordinates (c).

A linear system $\mathscr{L}$ is said to be *ample* if it has no fixed divisor, and if the rational map $T : V \to V'$ induced by $\mathscr{L}$ is biholomorphic at every point of V and V'. If D is a divisor such that $\mathscr{L}(D)$ is ample, then we say that D is *ample*. The system of hypersurface sections discussed previously is ample. If $\mathscr{L}$ is ample, then one sees immediately from the above discussion that a divisor X in $\mathscr{L}$ corresponds to a hyperplane section on V' under the biholomorphic correspondence $T : V \to V'$.

We conclude this section with a criterion for a linear system to be ample. Recall that a rational map may be everywhere defined and $1 - 1$ without being biholomorphic. For instance, the rational map from the affine line to the curve $y = t^2$, $x = t^3$ is everywhere defined and $1 - 1$, but not biholomorphic. The origin on the latter curve is a singular point, and its local ring is not equal to the local ring of the origin on the line. In addition to that, the rational map may be purely inseparable, and not birational.

We shall say that a linear system *separates points* if for each pair of points $P, Q \in V$, there exists $X \in \mathscr{L}$ such that $P \in \operatorname{supp}(X)$ but $Q \notin \operatorname{supp}(X)$.

PROPOSITION 7. *Let V be normal and let $\mathscr{L}$ be a linear system on V without fixed divisor. Then the associated rational map $T : V \to V'$ is everywhere defined and $1 - 1$ if and only if $\mathscr{L}$ separates points.*

Proof: Assume $\mathscr{L}$ separates points. Then $\mathscr{L}$ cannot have base points, and hence T is everywhere defined by the preceding proposition. Let P, Q be two distinct points of V. Say $P \in \operatorname{supp}(X_0)$ and $Q \notin \operatorname{supp}(X_0)$ for $X_0 \in \mathscr{L}$. Then we can dehomogenize $\mathscr{L}$ at X_0 to get a vector space L_0 of functions which are all defined at Q. If $f_0, \ldots, f_n$ is a basis for L_0, then $T(Q)$ is represented by $(f_0(Q), \ldots, f_n(Q))$. For some $Y \in \mathscr{L}$, $P \notin \operatorname{supp}(Y)$, and there is a function $g \in L_0$ such that $(g) = Y - X_0$, and g is therefore defined at P. Hence $T(P)$ cannot have a representative on the dehomogenized affine representative derived from L_0. This shows that $T(P) \neq T(Q)$, i.e. T is $1 - 1$. The converse is proved essentially by reversing the arguments.

As soon as we have a linear system which separates points, then we are not far off from finding one whose associated rational map T is an isomorphism. In fact, we can make use of the normalization of projective varieties to get a projective embedding of V if V is normal. If T is everywhere defined and $1 - 1$, then T is purely inseparable by the corollary of Proposition 5, Chapter IV, § 3. If we let V^* be the normalization of V' in the function field of V (which may be a finite extension of degree > 1 of the function field of V'), then we see immediately by ZMT that V and V^* are isomorphic. We shall give a refinement of this result in the next theorem.

THEOREM 4. *Let V be normal, and let D be a positive divisor on V. Let $T_n : V \to V'_n$ be the rational map induced by the linear system $\mathcal{L}(nD)$. If $\mathcal{L}(D)$ separates points, then for all large n, V'_n is projectively normal and T_n is biholomorphic between V and V'_n, so nD is ample.*

Proof: We shall first prove that for large n, T_n is birational. We know in any case that it is purely inseparable. All we shall use is that it is of finite degree, and the following lemma.

LEMMA. *Let D be a divisor $\geqq 0$ on an abstract variety V. If a function f on V satisfies an equation*

$$f^d + a_{d-1}f^{d-1} + \ldots + a_0 = 0$$

with $a_j \epsilon L(D)$, then $f \epsilon L(D)$.

Proof: If W is a pole of some coefficient in the above equation and if s is the maximum of the orders of the poles of the a_i at W, then f has a pole at W of order $\leqq s$. For otherwise, dividing the equation by f^d and applying the place corresponding to W would give $1 = 0$, a contradiction.

The union of all $L(nD)$ for $n = 1, 2, \ldots$, is the ring of all functions having poles only among the prime divisors occurring with multiplicity $\geqq 1$ in D. In view of the lemma, this ring is integrally closed in the function field of V, hence this function field is its quotient field, and is therefore the quotient field of one of the rings $k[L(n_0 D)]$ for some n_0, and hence for all $n > n_0$, as was to be shown.

We assume from now on that T is birational (in addition to being everywhere defined and $1 - 1$). Let $L = L(D)$ and let $f_0, \ldots, f_m$ be a basis for L. Let P be a generic point of V over k, and let $y_i = f_i(P)$. Let t be transcendental over $k(y)$. Then $R = k[ty]$ is a homogeneous coordinate ring for V'. Let I be its integral closure and let R_n, I_n be their homogeneous elements of degree n, following the notation of Chapter V. Let V_n^* be the projective variety whose homogeneous coordinate ring is $k[I_n]$. Then for all large n, we know from Theorem 6 of Chapter V, §5 that V_n^* is projectively normal, and that it is a normalization of V', hence is isomorphic to V'. Furthermore I_n gives rise to the linear system of hyperplane sections on V_n^*, and this linear system is complete by Theorem 3.

We contend that $I_n/t^n = L(nD)$ for all large n. Suppose this is proved. Noting that t^n is transcendental over the function field $k(y)$ of V, we see by definition that $k[I_n] = k[t^nL(nD)]$ is a homogeneous coordinate ring for V_n'. Showing that $I_n/t^n = L(nD)$ will therefore conclude the proof of our theorem.

We first show that $I_n/t^n \subset L(nD)$. Let $f \in I_n$. Then f satisfies an equation

$$f^d + a_{d-1}f^{d-1} + \ldots + a_0 = 0$$

with $a_i \in R$. We may assume that the coefficient a_i is homogeneous of degree $s_i = n(d - i)$, so that $s_i + ni = nd$. Dividing the equation by t^{nd} we see that the function $g = f/t^n$ satisfies an integral equation with coefficients in $L(nD)$. By the lemma, we conclude that $g \in L(nD)$.

Now we know that I_n/t^n is a space of functions on V_n^* defining the complete linear system of hyperplane sections. On the other hand, $L(nD)$ defines a complete linear system on V, and these two systems obviously correspond under the biholomorphic correspondence between V and V_n^*. Hence $\dim L(nD) = \dim I_n = \dim (I_n/t^n)$ where $\dim$ denotes the linear dimension over k. This concludes the proof of our theorem.

The same method as was used in Theorem 4 can also be used to give another criterion for a divisor to be ample.

PROPOSITION 8. *Let U be complete, normal, and let D be a positive* *divisor on U. Assume that $\mathscr{L}(D)$ has no base points. Let $T : U \to V$ be the rational map induced by $\mathscr{L}$, and assume that there is an integer d such that given any point Q of V there are at most d points in the inverse image $T^{-1}(Q)$. Then for all n sufficiently large, $\mathscr{L}(nD)$ is ample.*

Proof: Note that T is everywhere defined. The second hypothesis shows that $\dim V = \dim U$, and since U is assumed normal, it is the normalization of V in the function field of U, which is a finite extension of that of V. Let $f_0 = 1$, $f_1, \ldots, f_m$ be a basis for $L(D)$ over k. Then $k[f_0, \ldots, f_m]$ is the affine coordinate ring of an affine representative V_0 of V. The integral closure of this ring in $k(U)$ is finitely generated, and is a coordinate ring for the normalization V_0^* of V_0 in $k(U)$. The functions of $k(U) = k(V_0^*)$ which are integral over $k[f_0, \ldots, f_m]$ have poles on U only at the components of D (by the lemma). Hence there exists an integer e_0 such that $L(e_0 D)$ contains a set of generators for the coordinate ring of V_0^*. If we denote by $T^* : U \to V^*$ the rational map induced by $\mathscr{L}(e_0 D)$, it follows that T^* is birational, and is biholomorphic at every point P of U such that $T(P)$ has a representative on the affine variety V_0.

Now if we write $(f_i) = D_i - D$, we can dehomogenize $\mathscr{L}$ at the finite number of divisors $D_0 = D, \ldots, D_m$ and we obtain affine representatives $V_0, \ldots, V_m$ of V which cover V. We let $T_i : U \to V_i$ be the rational map induced by T on these affine models. For each i, we take an integer e_i which achieves with respect to V_i the same object that we achieved with e_0 with respect to V_0. Let e be an integer greater than any of the e_i, and let $T' : U \to V'$ be the rational map determined by the linear system $\mathscr{L}(eD)$. Then T' is biholomorphic at every point P of U such that $T(P)$ has a representative on at least one V_i. As this accounts for all points of U, we have shown that T' is everywhere biholomorphic. This proves our proposition.

5. Divisors rational over a field

In this section we prove two important theorems concerned with the rationality of divisors over a field k. The first one shows how to extend to arbitrary ground fields the theory of divisors and linear systems derived before over an algebraically closed field k. The second takes care of some unfinished business (Theorem 12 of Chapter III, § 4). We have followed closely the proofs given by Weil in F — VIII_3 Theorem 10 (better known as "the last theorem in *Foundations*"). An analysis of these proofs shows that they depend only on the results which we have given in Chapter III and some elementary statements concerning discrete valuations. They do not depend on everything that precedes them in *Foundations* (as a careless reader might have thought).

We begin by some results on the behavior of local rings under constant field extensions.

PROPOSITION 9. *Let* (x) *be the generic point of a variety* V *defined over a field* k. *Let* (x') *be a point of* V, *and let* E *be an extension of* k *which is free from* $k(x)$ *and* $k(x')$ *over* k. *Assume that* $k(x')$ *is separable over* k. *Let* $\mathfrak{o}_k$ *be the local ring of* (x') *in* $k(x)$, *and* $\mathfrak{o}_E$ *the local ring of* (x') *in* $E(x)$. *Let* $\mathfrak{m}_k$ *and* $\mathfrak{m}_E$ *be their maximal ideals. Then* $\mathfrak{m}_E = \mathfrak{m}_k \mathfrak{o}_E$.

Proof: We may assume without loss of generality that E is finitely generated over k. In that case, we can break up the extension E over k into three steps, purely transcendental, separable algebraic, and purely inseparable. If E is purely transcendental or purely inseparable over k, then it is linearly disjoint from $k(x)$ and $k(x')$ over k. Our proposition will then be a consequence of the following result.

PROPOSITION 10. *Let* (x) *be a generic point of* V *over* k, *and* (x') *a point of* V. *Let* E *be an extension of* k *which is linearly disjoint from* $k(x)$ *and* $k(x')$ *over* k. *Then* $\mathfrak{m}_E = \mathfrak{m}_k \mathfrak{o}_E$.

Proof: Let y be in $\mathfrak{m}_E$. We can write $y = f(x)/g(x)$, $f, g \in E[X]$ and $f(x') = 0$, but $g(x') \neq 0$. Since $1/g(x)$ is in $\mathfrak{o}_E$ all we need to do is prove that $f(x)$ is in $\mathfrak{m}_k\mathfrak{o}_E$. Writing $f = \sum w_\alpha f_\alpha(X)$ where $\{w_\alpha\}$ is a linear basis for E over k, we see from the linear disjointness of $k(x')$ and E that $f(x') = 0$ implies $f_\alpha(x') = 0$ for all α. Hence each $f_\alpha(x)$ is in $\mathfrak{m}_k$, and we are done.

We are left with the consideration of the case where E is finite separable algebraic over k.

Suppose that we have a quotient $f(x)/g(x)$ in $\mathfrak{o}_E$, with $f(x') = 0$, but $g(x') \neq 0$. It will suffice to prove our proposition for $f(x)$ in $E[x]$. Let A be the k-closed set consisting of all specializations of (x') over k (it is a k-subvariety of V). Remark right away that the ideal in $E[X]$ vanishing on A has a basis in $k[X]$, say, by virtue of Theorem 11 of Chapter III, § 4, because $k(x')$ is assumed separable over k, and hence the cycle consisting of all components of A each counted with multiplicity 1 is rational over k. Over E, we have a decomposition of A into E-components, and one of them, say W, will be the locus of (x') over E. We can write $A = W \cup B$, where B is an E-closed set all of whose E-components have the same dimension as W, and there is no inclusion relation between the E-components of A (so that $W \not\subset B$). There exists a polynomial $h(X)$ in $E[X]$ such that $h(x') \neq 0$ but h vanishes on B. (Select an element in the prime ideal belonging to each of the E-components of B, but not in the prime ideal of W in $E[X]$, and take the product.) Then $f(X)h(X)$ vanishes on A, and since the ideal of A in $E[X]$ has a basis in k, we can write $f(X)h(X) = \sum c_j f_j(X)$, where c_j is in E and $f_j(X)$ has coefficients in k and vanishes on A, hence on (x'). Dividing by $h(x)$, we see that $f(x) = \sum c_j f_j(x)/h(x)$ is in $\mathfrak{o}_E \mathfrak{m}_k$ because all the $f_j(x)$ are in $\mathfrak{m}_k$. This proves Proposition 9.

We shall apply our proposition to the case where V is of dimension r and where (x') is of dimension $r - 1$ over k. For the rest of this chapter, such a point will be said to be *simple on V* if its local ring is integrally closed, and hence is a discrete valuation ring. The justification for this terminology will be given in the chapter on simple points. A subvariety W of V of dimension $r - 1$ will be said to be *simple* if, whenever k is a field of definition for V such that W is defined over $\bar{k}$, a generic point of W over $\bar{k}$ is simple on V.

Suppose that we have an extension L of $K = k(x)$ and a valuation of L inducing a valuation of K. Let Γ_L and Γ_K be the value groups, as we have seen in Chapter I, § 6. If $\Gamma_L = \Gamma_K$, we say that the valuation of K is *unramified* in L. In the special case of discrete valuations, this means that a generator for the value group of Γ_L can be selected in Γ_K, i.e. can be selected to be the value of an element of K.

In Proposition 10, let (x') have dimension $r-1$ over k, and assume (x') is simple on V. Then by hypothesis, its local ring in $k(x)$ is a discrete valuation ring, and an element of order 1 in that ring also has order 1 in the value group associated with the valuation ring of (x') in $E(x)$. In other words, the valuation is unramified. This comes from the fact that the maximal ideal $\mathfrak{m}_k$ has one generator which can be used as a generator of $\mathfrak{m}_E$. In particular, if k is a field of definition for a subvariety W of V, of dimension $r-1$, and if (x') is a generic point of W over k, then we can use the discrete valuation in $k(x)$ arising from $\mathfrak{o}_k$ in order to compute the order of a function in $k(x)$ at W.

Now let V be an abstract variety. Let f be a function on V, i.e. a rational map of V into S^1, and let W be a simple subvariety of V, of dimension $r-1$. We wish to define the order of f at W. We take any field of definition k for V, and W. Without loss of generality, we may assume that V is affine. We take a generic point (x) of V over k, and a generic point (x') of W over k. By hypothesis, the local ring of (x') in $k(x)$ is a discrete valuation ring. The order of $f(x)$ under this valuation defines $\mathrm{ord}_W f$. Proposition 10 shows that this integer does not depend on the field of definition for f, V, and W selected. We say that f has a zero or a pole at W according as $\mathrm{ord}_W f$ is > 0 or < 0. If W is not simple on V, then we agree to define $\mathrm{ord}_W f = 0$.

PROPOSITION 11. *Let V be an abstract variety, and f a function on V, i.e. a rational map of V into S^1. Let T be the graph of f. Then the zeros of f are the simple subvarieties of V appearing in the projection on V of $T \cap (V \times 0)$.*

Proof: Using the dimension theorem, it is easily seen that each component of this intersection has dimension $r-1$. The proposition is then trivial from the definitions.

The above proposition gives a geometric description for the zeros of a function. For the poles, we can of course consider $1/f$ instead of f. If we had intersection theory, we could give a geometric definition for the multiplicities $\mathrm{ord}_W f$. From the proposition, we also get a corollary.

COROLLARY. *Let f be a function on abstract variety, V, and k, a field of definition for f and V. Let W be a subvariety of V of codimension 1, and let E be the smallest field of definition for W containing k. If E is not algebraic over k, then $\mathrm{ord}_W f = 0$.*

Proof: This is immediate from the theory of affine varieties.

If W is defined over $\bar{k}$, it may not be defined over a separable extension of k, and it will now be necessary to investigate how the discrete valuation behaves in that case.

Let V be an affine variety defined over k, and let (x') be a simple point of V, of dimension $r - 1$ over k. Let $\mathfrak{o}_k$ be as before the local ring of (x') in $k(x)$, where (x) is a generic point of V over k. Then by hypothesis, $\mathfrak{o}_k$ is a discrete valuation ring. Let p^m be the order of inseparability of $k(x')$ over k and let $E = k^{1/p^m}$. According to $\mathrm{F} - \mathrm{I}_8$ Proposition 31, $E(x)$ is separable over E. We have a discrete valuation ring $\mathfrak{o}_E$ for (x') in $E(x)$. Let v_k and v_E be the valuations. Then we shall prove that the ramification index e divides p^m. Indeed, every element u of $E(x)$ is such that u^{p^m} lies in $k(x)$. Let u be a generator of $\mathfrak{m}_E$, the maximal ideal of $\mathfrak{o}_E$. Then u^{p^m} is in $\mathfrak{m}_k$, the maximal ideal of $\mathfrak{o}_k$. If t is a generator of $\mathfrak{m}_k$ as an ideal of $\mathfrak{o}_k$, then for some power s of t, the ideal (u^{p^m}) of $\mathfrak{o}_k$ is equal to (t^s). On the other hand, the ideal (t) of $\mathfrak{o}_E$ is equal to (u^e). This proves what we wanted.

We see that our results be can expressed in geometric language in the following manner.

PROPOSITION 12. *Let V^r be an abstract variety defined over a field k. Let W be a simple subvariety of codimension 1 on V, defined over $\bar{k}$, and having the order of inseparability p^m over k. Then there is a function f on V, defined over k, such that $\mathrm{ord}_W f = p^m$.*

Proof: Without loss of generality, we may assume that V is an affine variety. Let (x) be a generic point of V over k, and (x') a generic point of W over $\bar{k}$. The order of inseparability of W over k is by definition that of $k(x')$ over k. Going from k to k^{1/p^m} we get a ramification index dividing p^m. After that, we may take a separable algebraic extension of k^{1/p^m} over which W is defined, and we know that the discrete valuation from (x') in $E(x)$ is unramified in such a separable extension. This shows that a generator t of $\mathfrak{m}_k$ which can be written $t = f(x)$ gives rise to a function f having order dividing p^m at W. This proves our proposition.

REMARK. It can be shown that the ramification index is exactly equal to the order of inseparability p^m. We leave this as an exercise to the reader.

Let V be an abstract variety defined over a field k, and let D be a divisor on V, rational over k. We shall denote by $L_k(D)$ the k-vector space of functions f defined over k such that $(f) \geq -D$, and by $M_k(D)$ the k-vector space of functions f defined over k such that $(f) \geq D$. We have $M_k(D) = L_k(-D)$. Each one is a subspace of $k(V)$ viewed as vector space over k. Let E be a field containing k. If A is a subspace of $k(V)$ over k, we shall denote by $E\langle A \rangle$ the subspace of $E(V)$ generated by A over E, i.e. consisting of linear combinations of elements of A with coefficients in E (or symmetrically, of linear combinations of elements of E with coefficients in A). Our first main theorem states that $M_E(D)$ consists precisely of the space obtained by lifting $M_k(D)$ over E. In other words, we shall prove:

THEOREM 5. *Let V be an abstract variety defined over a field k. Let D be a divisor on V, rational over k. Let E be a field containing k. Then $M_E(D) = E\langle M_k(D) \rangle$.*

Proof: Without loss of generality, we may assume E finitely generated over k. Furthermore, if $E \supset F \supset k$ is a tower of fields, and if we have proved our theorem in each step of the tower, i.e. if we have proved $M_E(D) = E\langle M_F(D) \rangle$ and $M_F(D) = F\langle M_k(D) \rangle$, then obviously $M_E(D) = E\langle M_k(D) \rangle$. Thus it will suffice to prove our theorem in case E is generated by one element which is separable algebraic, purely transcendental, or purely inseparable over k.

Case 1. E is separable algebraic over k.

In this case, the composite field $k(V)E$ is the same as the vector space $E\langle k(V)\rangle$, and is precisely the field $E(V)$. Thus each element of that field can be written $f = \sum c_j f_j$, where the c_j are linearly independent elements of E over k and $f_j \in k(V)$. The field $k(c)$ being separable algebraic, it follows from the theory of separable algebraic extensions that when σ ranges over all distinct isomorphisms of $k(c)$ over k, then the determinant det (c_j^σ) is not 0. Applying each σ to the expression $f = \sum c_j f_j$ we get $f^\sigma = \sum c_j^\sigma f_j$ because f_j is in $k(V)$ and is fixed under σ. We can now solve for each f_j as a linear combination of the f^σ with coefficients in some extension F of k. By first principles, we note that $(f) \geqq D$ implies $(f^\sigma) \geqq D^\sigma$ and hence $(f^\sigma) \geqq D$ because D is rational over k and hence fixed under σ. Since $M_F(D)$ is a vector space we conclude that $(f_j) \geqq D$, as desired.

Case 2. E is purely transcendental over k.

The proof in this case will follow the same pattern as in the preceding case, but there is one added difficulty. It is of course not true any more that $E\langle k(V)\rangle$ is equal to the field $E(V)$ and hence we shall first have to prove that a function $f \in E(V)$ satisfying $(f) \geqq D$ with D rational over k is in fact in $E\langle k(V)\rangle$.

Suppose first that V is affine, and let (x) be a generic point of V over E, and hence over k. Let $K = k(x)$, and $EK = K(t)$ where t is transcendental over k and $E = k(t)$. Let $f(x) = z$. It is an element of $K(t)$, which can be written

$$\frac{g(t)}{h(t)} = \frac{a_n t^n + \ldots + a_0}{t^m + b_{m-1} t^{m-1} + \ldots + b_0}$$

with $g(t)$, $h(t)$ in $K[t]$, and relatively prime in $K[t]$. If all coefficients $b_j \in k$, then we see that $z \in E\langle K\rangle$, and we are done.

Otherwise, we shall prove that f has a pole at a subvariety W of V which is not defined over k, and shall thus contradict the hypothesis that $(f) \geqq D$.

Suppose that not all b_j are in k. Then some irreducible factor $h_1(t)$ of $h(t)$ in $K[t]$ has a root u which is transcendental over k but algebraic over K, and $g(u) \neq 0$ because g and h are relatively prime in $K[t]$. There exists a specialization $(t, 1/z) \to (u, 0)$ over $k(x)$, and hence a specialization $(x, t, 1/z) \to (x, u, 0)$ over k.

The isomorphism $k[u] \to k[t]$ mapping u on t can be extended to an isomorphism of $k[u, x]$ which will map (x) on some point (x'). We then get a specialization (over k)

$$(x, t, 1/z) \to (x', t, 0)$$

by composing the two preceding specializations. The dimension of (x') over $k(t)$ is obviously $r - 1$, and since (x') is a generic specialization of (x) over k, the subvariety W of V having (x') as generic point over $k(t)$ cannot be defined over $\bar{k}$. It is a consequence of the theory of simple points that since (x') is a generic point of V over k, then it is simple when viewed as a generic point of W in the sense of this chapter. This shows that if f is not in $E\langle k(V) \rangle$ then f has a pole at the subvariety W of V above, a contradiction.

Now if V is an abstract variety and V_0 an affine representative defined over k, with a proper k-closed set F_0, then the variety W constructed above cannot be contained in F_0 because all components of F_0 are defined over $\bar{k}$. Hence the above arguments apply to abstract varieties as well.

The rest of the proof in Case 2 runs parallel to that of Case 1 once we have proved the following lemma.

LEMMA. *Let* $(w) = (w_1, \ldots, w_n)$ *be a point such that the extension* $k(w)$ *of* k *is regular, and such that the* w_j *are linearly independent over* k. *Let* $(w^{(1)}), \ldots, (w^{(n)})$ *be* n *generic independent specializations of* (w) *over* k. *Then the determinant* $\det (w_j^{(i)})$ *is not* 0.

Proof: Expand the determinant

$$\begin{vmatrix} w_1^{(1)} & \ldots & w_n^{(1)} \\ \ldots & \ldots & \ldots \\ w_1^{(n)} & \ldots & w_n^{(n)} \end{vmatrix}$$

according to the first row. If the determinant is 0, then this gives a linear combination of $w_1^{(1)}, \ldots, w_n^{(1)}$ with coefficients in the field generated over k by the other $(w^{(\nu)})$. Since we assume $w_1, \ldots, w_n$ linearly independent over k, an induction leads to a contradiction of Theorem 3 of Chapter III, § 2.

Returning to the proof of Case 2, we can write $f = \sum c_j f_j$ with

$f_j \epsilon k(V)$ and c_j linearly independent elements of $k(t)$ over k. We now use the lemma, and let σ range over n isomorphisms of $k(c)$ mapping (c) on n independent generic specializations of (c). We can then solve for each f_j as in Case 1, and this concludes the proof of Case 2.

Case 3. E is purely inseparable over k.

The technique for this case of course differs from the other two. Since E is algebraic over k, we have the same advantage as in Case 1 that $E(V)$ is actually equal to $E\langle k(V)\rangle$.   Hence our function f can be written $f = \sum c_j f_j$, where $f_j \epsilon k(V)$, and the c_j are elements of E linearly independent over k. Let W be a component of D, and p^m its order of inseparability over k. If σ is an automorphism of $\bar{k}$ over k, then σ leaves f fixed (because f is defined over the purely inseparable extension E of k) and by hypothesis we have $\mathrm{ord}_{W\sigma}f \geqq sp^m$, where sp^m in the coefficient of W in D. It will suffice to prove that each f_j satisfies the same condition, i.e. $\mathrm{ord}_{W\sigma}f_j \geqq sp^m$. If W has an affine representative on some affine representative V_0 of V, then each W^σ has also a representative on V_0 (because the frontier on V_0 is k-closed).   Hence we may assume without loss of generality that we deal with an affine variety from the beginning.

Let V be affine and let g be a function in $k(V)$ such that $\mathrm{ord}_W g = p^m$ (such a function exists by Proposition 12). Then we get

$$\mathrm{ord}_W(g^{-s}f) = \mathrm{ord}_W f - s \cdot \mathrm{ord}_W g \geqq 0.$$

Applying an automorphism of $\bar{k}$ over k, using first principles, and noting that $f^\sigma = f$ for every such σ because E is purely inseparable over k, we see that $g^{-s}f$ has no pole at W or any one of the conjugates W^σ of W over k.

Let (x) be a generic point of V over E, and hence over k. Put $t = g(x)$, $y = f(x)$ and $y_j = f_j(x)$. Let (x') be a generic point of W over $\bar{k}$. Then the preceding paragraphs implies that $t^{-s}y$ lies in the local ring of (x') in $E(x)$. Hence we can write

$$\sum c_j(y_j t^{-s}) = \frac{P(x)}{Q(x)},$$

where $P(x)$ and $Q(x)$ are in $E[x]$ and $Q(x') \neq 0$. If p^n is the degree of E over k, multiply the above equation by $Q(x)^{p^n}$. We get

$$P(x)Q(x)^{p^n-1} = \sum c_j (y_j t^{-s} Q(x)^{p^n}).$$

The expression on the left lies in $E[x]$. The quantities t^{-s}, y_j and $Q(x)^{p^n}$ lie in $k(x)$. Furthermore, $k(x)$ and E are linearly disjoint because $k(x)$ is regular over k (Theorem 3 of Chapter III, § 2). We can complete the $\{c_j\}$ to a basis of E over k and every element of $E[x]$ can be written as a linear combination of the c_j with coefficients in $k[x]$. In view of the linear disjointness, these coefficients are unique. This proves that $y_j t^{-s} Q(x)^{p^n}$ is actually in $k[x]$. We can therefore find $h_j(x)$ in $k[x]$ such that

$$y_j = t^s h_j(x)/Q(x)^{p^n}.$$

Since $Q(x')^{p^n} \neq 0$, and since the function h_j in $k(V)$ defined by $h_j(x)$ obviously has no pole at W, we conclude from the definitions that $\mathrm{ord}_W f_j \geqq s p^m$, as was to be shown.

As a corollary, we state explicitly the form given by Weil to the theorem.

COROLLARY 1. *Let V be an abstract variety defined over a field k. Let D be a divisor on V, rational over k. Let f be a function on V, other than the constant 0, and let E be a field of definition for f containing k. Then, if $(f) \geqq D$, f can be written in the form $f = \sum c_j f_j$ where the c_j are linearly independent elements of E over k, and f_j are functions on V, other than the constant 0, having k as field of definition When f is so expressed, we have $(f_j) \geqq D$ for every j.*

Proof: The existence follows from the theorem. Since $k(V)$ is isomorphic to $k(P)$, where P is a generic point for V over the field $k(c)$, it follows that $k(V)$ and $k(c)$ are linearly disjoint over k, and the f_j are then uniquely determined. From the theorem, we can write $f = \sum b_i g_i$, where b_i is part of a basis of $k(c)$ over k, and $(g_i) \geqq D$. The f_j are linear combinations of the g_i with constant coefficients. Hence they also satisfy the condition $(f_j) \geqq D$.

COROLLARY 2. *Let V be a complete variety, having no singular* *subvarieties of codimension* 1. *Let D be a divisor on V, rational over a field of definition k of V. If there exists a function f on V such that* $(f) = D$, *then there exists a function g on V defined over k, such that* $(g) = D$.

Proof: One considers the space $L_k(D)$ and uses the fact that a function without zero or pole is a constant.

COROLLARY 3. *Let U be a variety, complete and normal, and defined over a field k. Assume that U has a projective embedding over $\bar{k}$. Then U has a projective embedding over k, i.e. there exists a complete linear system $\mathscr{L}$ which contains a divisor rational over k, and whose induced rational map $T : U \to V$ is birational and biholomorphic.*

Proof: We can find by hypothesis a divisor D on U, rational over a finite extension k' of k, such that $\mathscr{L}(D)$ is ample. Let $D_1, \ldots, D_m$ be the distinct conjugates of D over k, and let p^s be a sufficiently high power of the characteristic, such that $p^s \sum D_i$ is rational over k. Put $E = p^s \sum D_i$. Since $\mathscr{L}(D)$ separates points, it is clear $\mathscr{L}(E)$ separates points. Furthermore, since E is rational over k, $L(nE)$ has a basis of functions defined over k. Hence the rational map of U onto a projective variety derived from $\mathscr{L}(nE)$ is defined over k. Since we know that for all large n, $\mathscr{L}(nE)$ is ample, it follows that we get a projective embedding for U defined over k for all large n.

For the analogous result when U is not assumed complete, or normal, we refer the reader to Weil's article. "The field of definition of a variety," American Journal of Mathematics, Vol. 78 No. 3, July 1956, pp. 509—524, Theorem 7.

Our second theorem will be proved by using the first. We make some preliminary remarks on affine varieties. Let V be an affine variety defined over a field k. In the coordinate ring $\Omega[X]$ over the universal domain, a polynomial $f(X)$ determines a function on V in the following manner. Let K be a field containing k and all coefficients of f. Select a generic point (x) of V over K. Then $f(x)$ is in $K[x]$, and determines a function in $K(V)$. This function is clearly independent of the field K selected subject to the above conditions. Topologically speaking, it is the restriction to V of the polynomial function $f(X)$ on affine space. The divisor of the function thus induced by $f(X)$ on V will be denoted by $(f)_V$. This being the case, we can now state our theorem as follows.

THEOREM 6. *Let V be an abstract variety defined over a field k. Let D be a divisor on V. Then there exists a smallest field K containing k over which D is rational. If V is affine, if $D \geqq 0$, and if all components of D occur with the same multiplicity, then this smallest field K is the smallest field of definition of the ideal $\mathfrak{a}$ in $\Omega[X]$ consisting of all polynomials $f(X)$ which either vanish on V or are such that $(f)_V \geqq D$.*

Proof: Without loss of generality, we may assume immediately that V is affine. Write $D = \sum mD_m$, where D_m is the sum of the terms in D appearing with coefficient m. Then D is rational over a field if and only if each mD_m is rational over that field. Hence it will suffice to deal with a divisor of type mD_m. Furthermore, if we write $m = p^s m'$, where m' is prime to p, then mD_m is rational over K if and only if $p^s D_m$ is rational over K. Hence it will suffice to deal with a divisor D which can be written

$$D = qD_0,$$

where q is a power of the characteristic $p \neq 0$, and D_0 is a divisor on V which is the sum of distinct components, appearing with coefficient 1 in D_0.

If D is rational over a field K containing k then it is an immediate consequence of Theorem 5 that K is a field of definition for $\mathfrak{a}$.

Conversely, assume that K is a field of definition for $\mathfrak{a}$ containing k. We shall prove that D is rational over K. The existence of a smallest field of definition for an ideal will then finish the proof of our theorem.

Remark that a polynomial $f(X)$ is 0 on the support of D if and only if some power f^v lies in $\mathfrak{a}$. Indeed, if f vanishes on supp (D), then $(f)_V$, which consists of components having only positive coefficients, has all the components of D with non-zero coefficient. Hence a suitable power of f lies in $\mathfrak{a}$ by definition. The converse is equally clear. This remark implies that the support of D is K-closed and that D_0 is rational over $K' = K^{1/p^\infty}$. Let W be a component of D, and let E be a field of definition for all the components of D (or D_0) and containing K'. By Proposition 5, there exists a polynomial $f(X)$ such that

$$(f)_V = D_0 + B,$$

where B is a divisor having no components in common with D_0. According to Theorem 5, we can write

$$f = \sum c_j f_j,$$

where the $\{c_j\}$ are linearly independent elements of E over K', and the f_j are in $K'[X]$, and in that case $(f_j)_V \geqq D_0$ for all j. Furthermore, W must have the coefficient 1 in at least one of the f_j, since otherwise it would occur in B. If we call that polynomial g, then g is in $K'[X]$, and W has coefficient 1 in $(g)_V$, and we have $(g)_V \geqq D_0$. But then g^q is in $\mathfrak{a}$ by definition, and therefore, by hypothesis, may be written $\sum c_\alpha g_\alpha$, where g_α are in $\mathfrak{a} \cap K[X]$. This latter fact implies that W has at least the coefficient q in all $(g_\alpha)_V$. As it has the coefficient q in $(g^q)_V$, it must have the coefficient q in at least one of the divisors $(g_\alpha)_V$. As these divisors are rational over K, this implies that if D_1 is the sum of W and its conjugates over K, the divisor qD_1 is rational over K. As this is so for every component W of D, D is therefore rational over K, and this concludes the proof.

Literature

The theorem concerning the finite dimensionality of a complete linear system on an abstract, complete variety is a special case of a theorem on coherent sheaves proved by Serre in [6]. This more general result yields for instance an analogous statement for differential forms.

Theorems giving more precise information concerning the dimension $l(D)$ are said to be of Riemann-Roch type. In this connection, we refer the reader to Zariski [8], and for a cohomological approach to Serre [5]. In the classical case, one has of course the extensive results of Hirzebruch [2], Kodaira and Spencer [4] [7].

Recent proofs that certain varieties have a projective embedding depend on the study of suitable linear systems. The literature is extensive, and so we give below only two papers dealing with such questions: Chow's proof that homogeneous spaces have a projective embedding, and Kodaira's paper on the embedding of

Hodge manifolds [1], [3]. From these papers, the reader will have no difficulty in tracing the rest of the literature.

CHOW, W. L.

[1] Projective embedding of homogeneous spaces, Lefschetz conference volume, Princeton University Press, 1957.

HIRZEBRUCH, F.

[2] Neue topologische Methoden in der algebraischen Geometrie, Springer, 1956.

KODAIRA, K.

[3] On Kahler varieties of restricted type, Annals of Math., Vol. 60, No. 1, July 1954, pp. 28—48.

KODAIRA-SPENCER.

[4] On arithmetic genera of algebraic varieties, Proc. of the Nat. Acad. of Sciences, Vol. 39, No. 7, July 1953, pp. 641—649.

SERRE, J. P.

[5] Faisceaux algébriques cohérents, Annals of Math., Vol. 61, No. 2, March 1955, pp. 197—278.

[6] Sur la cohomologie des variétés algébriques, Journ. math. pures, 36, 1957, pp. 1—16.

SPENCER, D. C.

[7] Cohomology and the Riemann Roch theorem, Proc. of the Nat. Acad. of Sciences, Vol. 39, No. 7, July 1953, pp. 660—669.

ZARISKI, O.

[8] Complete linear systems on normal varieties and a generalization of a lemma of Enriques Severi, Annals of Math., Vol. 55, No. 3, May 1952, pp. 552—592.

CHAPTER VII

Differential Forms

We reproduce here parts of the contents of Koizumi's paper "On the differential forms of first kind on algebraic varieties," J. Math. Soc., Japan, 1949. The main result is the birational invariance of the differential form of first kind, i.e. those which are everywhere holomorphic. We work over an algebraically closed field k, and we leave it to the reader to prove the theorem analogous to Theorem 5 of Chapter VI, for differential forms.

1. Derivations

A *derivation* D of a ring R is a mapping $D : R \to R$ of R into itself which is linear and satisfies the ordinary rule for derivatives, i.e. $D(x + y) = Dx + Dy$, and $D(xy) = xDy + yDx$. As an example of derivations, consider the polynomial ring $k[X]$ over a field k. For each variable X_i, the partial derivative $\partial/\partial X_i$ taken in the usual manner is a derivation of $k[X]$. We also get a derivation of the quotient field in the obvious manner, i.e. by defining $D(u/v) = (vDu - uDv)/v^2$.

We shall work with derivations of a field K in this chapter. A derivation of K is *trivial* if $Dx = 0$ for all $x \epsilon K$. It is trivial *over a subfield* k of K if $Dx = 0$ for all $x \epsilon k$. A derivation is always trivial over the prime field: one sees that $D(1) = D(1 \cdot 1) = 2D(1)$, whence $D(1) = 0$.

We now consider the problem of extending derivations. Let $L = K(x) = K(x_1, \ldots, x_n)$ be a finitely generated extension. If $f \epsilon K[X]$, we denote by $\partial f/\partial x_i$ the polynomials $\partial f/\partial X_i$ evaluated at (x). Given a derivation D on K, does there exist a derivation D^* on L coinciding with D on K? If $f(X) \epsilon K[X]$ is a polynomial vanishing on (x), then any such D^* must satisfy

$$(1) \qquad 0 = D^*f(x) = f^D(x) + \sum (\partial f/\partial x_i)D^*x_i.$$

where f^D denotes the polynomial obtained by applying D to all coefficients of f. Note that if relation (1) is satisfied for every element in a finite basis of the ideal in $K[X]$ vanishing on (x), then (1) is satisfied by every polynomial of this ideal. This is an immediate consequence of the rules for derivations.

The above necessary condition for the existence of a D^* turns out to be sufficient.

THEOREM 1. *Let D be a derivation of a field K. Let (x) be any set of quantities, and let $f_\alpha(X)$ be a basis for the ideal determined by (x) in $K[X]$. Then, if (u) is any set of elements of $K(x)$ satsifying the equations*

$$0 = f_\alpha^D(x) + \sum (\partial f_\alpha / \partial x_i) u_i,$$

there is one and only one derivation D^* of $K(x)$ coinciding with D on K, and such that $D^*x_i = u_i$ for every i.

Proof: The necessity has been shown above. Conversely, if $g(x)$, $h(x)$ are in $K[x]$, and $h(x) \neq 0$, one verifies immediately that the mapping D^* defined by the formulas

$$D^*g(x) = g^D(x) + \sum \frac{\partial g}{\partial x_i} u_i$$

$$D^*(g/h) = \frac{hD^*g - gD^*h}{h^2}$$

is well defined and is a derivation of $K(x)$.

Consider the special case where (x) consists of one element x. Let D be a given derivation on K.

Case 1. x is separable algebraic over K. Let $f(X)$ be the irreducible polynomial satisfied by x over K. Then $f'(x) \neq 0$. We have $0 = f^D(x) + f'(x)u$, whence $u = -f^D(x)/f'(x)$. Hence D extends to $K(x)$ uniquely. If D is trivial on K, then D is trivial on $K(x)$.

Case 2. x is transcendental over K. Then D extends, and u can be selected arbitrarily in $K(x)$.

Case 3. x is purely inseparable over K, so $x^p - a = 0$, with $a \in K$. Then D extends to $K(x)$ if and only if $Da = 0$. In particular if D is trivial on K, then u can be selected arbitrarily.

THEOREM 2. *A finitely generated extension $K(x)$ over K is separably algebraic if and only if every derivation D of $K(x)$ which is trivial on K is trivial on $K(x)$.*

Proof: If $K(x)$ is separable algebraic over K, this is Case 1. Conversely, if it is not, we can make a tower of extensions between K and $K(x)$, such that each step is covered by one of the three above cases. At least one step will be covered by Case 2 or 3. Taking the uppermost step of this latter type, one sees immediately how to construct a derivation trivial on the bottom and non-trivial on top of the tower.

THEOREM 3. *Given K and a point $(x) = (x_1, \ldots, x_n)$, assume that there exist n polynomials $f_i \in K[X]$ such that*
1. *$f_i(x) = 0$, and*
2. *det $(\partial f_i / \partial x_j) \neq 0$.*
Then (x) is separably algebraic over K.

Proof: Let D be a derivation on $K(x)$, trivial on K. Having $f_i(x) = 0$ we must have $Df_i(x) = 0$, whence the Dx_i satisfy n linear equations such that the coefficient matrix has non-zero determinant. Hence $Dx_i = 0$, so D is trivial on $K(x)$. Hence $K(x)$ is separable algebraic over K.

The following proposition will follow directly from Cases 1 and 2 above.

PROPOSITION 1. *Let $K = k(x)$ be a finitely generated extension of k. An element z of K is in $K^p k$ if and only if every derivation D of K over k is such that $Dz = 0$.*

Proof: If z is in $K^p k$, then it is obvious that every derivation D of K over k vanishes on z. Conversely, if $z \notin K^p k$, then z is purely inseparable over $K^p k$, and by Case 3 of the extension theorem, we can find a derivation D trivial on $K^p k$ such that $Dz = 1$. This derivation is at first defined on the field $K^p k(z)$. One can extend it to K as follows. Suppose there is an element $w \in K$ such that $w \notin K^p k(z)$. Then $w^p \in K^p k$, and D vanishes on w^p. We can then again apply Case 3 to extend D from $K^p k(z)$ to $K^p k(z, w)$. Proceeding stepwise, we finally reach K, thus proving our proposition.

The derivations D of a field K form a vector space over K if we define zD for $z \in K$ by $(zD)(x) = zDx$.

Let K be a finitely generated extension ot k, of dimension r over k. We denote by $\mathscr{D}$ the K-vector space of derivations D of K over k, (derivations of K which are trivial on k). For each $z \in K$, we have a pairing

$$(D, z) \to Dz$$

of $(\mathscr{D}, K)$ into K. Each element z of K gives therefore a K-linear functional of $\mathscr{D}$. This functional is denoted by dz. We have

$$d(yz) = ydz + zdy$$
$$d(y + z) = dy + dz.$$

These linear functionals form a subspace $\mathscr{F}$ of the dual space of $\mathscr{D}$, if we define ydz by

$$(D, ydz) \to yDz.$$

PROPOSITION 2. *Assume that K is a separably generated and finitely generated extension of k. Then the vector space $\mathscr{D}$ (over K) of derivations D of K over k had dimension r. Elements $t_1, \ldots, t_r$ of K form a separating transcendence base of K over k if and only if $dt_1, \ldots, dt_r$ form a basis of the dual space over K.*

Proof: If $t_1, \ldots, t_r$ is a separating transcendence base for K over k, then we can find derivations $D_1, \ldots, D_r$ of K over k such that $D_i t_j = \delta_{ij}$, by Cases 1 and 2 of the extension theorem. Given $D \in \mathscr{D}$, let $w_i = Dt_i$. Then clearly $D = \sum w_i D_i$, and so the D_i form a basis for $\mathscr{D}$ over K, and the dt_i are the dual basis. Conversely, if $dt_1, \ldots, dt_r$ is a basis for $\mathscr{F}$ over K, and if K is not separably generated over $k(t)$, then by Cases 2 and 3 above we can find a derivation D which is trivial on $k(t)$ but non-trivial on K. If $D_1, \ldots, D_r$ is the dual basis of $dt_1, \ldots, dt_r$ (so $D_i t_j = \delta_{ij}$) then $D, D_1, \ldots, D_r$ would be linearly independent over K, contradicting the first part of the theorem.

COROLLARY. *Let K be a finitely generated and separably generated extension of k. Let z be an element of K transcendental over k. Then K is separable over $k(z)$ if and only if there exists a derivation D of K over k such that $Dz \neq 0$.*

Proof: If K is separable over $k(z)$, then z can be completed to a separating base of K over k and we can apply the proposition. If $Dz \neq 0$, then $dz \neq 0$, and we can complete dz to a basis of $\mathscr{F}$ over K. Again from the proposition, it follows that K will be separable over $k(z)$.

2. Differential forms

We now apply the preceding section to varieties. We assume that k is algebraically closed, and that K is a finitely generated extension of k, and so can be interpreted as a function field. We denote again by $\mathscr{D}$ the space of derivations of K over k, and by $\mathscr{F}$ the dual space, and we recall that K is separable over k (Corollary 3 of Theorem 1, Chapter III, § 1.)

The Grassman algebra generated by $\mathscr{F}$ over K gives rise to the *differential forms* of K over k. A differential form of degree s is a homogeneous element of degree s of this algebra. Any differential form of degree s can be written

$$\omega = \sum_{i_1 < \ldots < i_s} z_{i_1} \ldots {}_{i_s} \, dt_{i_1} \wedge \ldots \wedge dt_{i_s} = \sum z_{(i)} dt_{(i)}.$$

If $s = 1$, and z is an element of K we can write

$$dz = \sum z_i dt_i$$

and the z_i are uniquely determined. We denote them by $\partial z / \partial t_i$.

We shall now study the local behavior of differential forms at simple points. For our purposes, we take the following property as a definition. Let V be an affine variety, a model of the finitely generated extension K of k of dimension r. We can write $K = k(x)$, where (x) is a generic point of V over k. A point (x') of V is said to be *simple* if there exist elements $t_1, \ldots, t_r$ in the local ring of (x') in $k(x)$, and polynomials $f_1(T, X), \ldots, f_n(T, X)$ in $k[T, X]$ such that

1. $f_i(t, x) = 0$, and
2. $\det (\partial f_i / \partial x'_j) \neq 0$.

The t_i being as above, we shall call them *local parameters* or *coordinates* of V at (x'). By Theorem 3, local coordinates form a separating transcendence base of $k(x)$ over k. A subvariety W of V is said to be *simple* on V if a generic point of W over k is simple on V. We note that if $(x') \to (x'')$ is a specialization, and (t) are local coordinates at (x''), then they are also local coordinates at (x'). In addition, any separating transcendence base (t) of K over k can be used as local coordinates for all points not lying on some algebraic set. Indeed, by Theorem 3 of § 1 we can use (t) as a set of local coordinates at the generic point (x), because the determinant above does not vanish for $(x') = (x)$. For all specializations $(x, t) \to (x', t')$ where (x') does not lie in some algebraic set, the determinant will not vanish.

PROPOSITION 3. *Let $(t_1, \ldots, t_r) = (t)$ be a set of local coordinates at a simple point (x') of V, and assume that z lies in the local ring of (x') in $k(x)$. If we write $dz = \sum z_i dt_i$ then each $z_i = \partial z / \partial t_i$ also lies in the local ring.*

Proof: Assume first $z = x_j$ is a coordinate of the generic point (x). If $f_i(T, X)$ are the polynomials satisfying conditions 1 and 2, then

$$\sum_{j=1}^{n} \frac{\partial f_i}{\partial x_j} dx_j + \sum_{\mu=1}^{r} \frac{\partial f_i}{\partial t_\mu} dt_\mu = 0.$$

Hence we can find polynomials $G_{j\mu}(T, X)$ such that

$$dx_j = \frac{\sum G_{j\mu}(t, x) dt_\mu}{\det(\partial f_i / \partial x_j)}.$$

Since the determinant does not vanish at (x'), we have proved Proposition 3 for each x_j. Now if $z = g(x)/k(x)$, and $h(x') \neq 0$, then

$$dz = \frac{h(x) dg(x) - g(x) dh(x)}{h(x)^2}$$

and the rest of the proof is clear.

If ω is a differential form, and if there exists a set of local coordinates (t) of V at a point (x') such that ω can be written

$$\omega = \sum z_{i_1 \cdots i_s} dt_{i_1} \wedge \ldots \wedge dt_{i_s} = \sum z_{(i)} dt_{(i)}$$

where all coefficients $z_{(i)}$ lie in the local ring of (x') in $k(x)$ (i.e. all $z_{(i)}$ are defined at (x')), then we shall say that ω is *holomorphic* at (x'). If (u) is another set of local coordinates at (x'), and ω is written $\sum y_{(i)} du_{(i)}$ and if all $z_{(i)}$ lie in the local ring of (x'), then it follows from Proposition 3 that all $y_{(i)}$ also lie in this ring. Hence our definition does not depend on the choice of parameters used to express ω, i.e. it depends only on ω and the point (x').

If the point (x') has dimension s over k, then it can be viewed as a generic point of a subvariety A of V of dimension s. If ω is holomorphic at (x'), then we shall also say that ω is *holomorphic at A*. This has special significance, whence $s = r - 1$ as shown by the following result.

THEOREM 4. *Let V be a normal variety defined over k. If the differential form ω is holomorphic at every subvariety W of V of codimension 1, then it is holomorphic at every simple point of V.*

Proof: Let (x') be simple on V, and suppose ω is not holomorphic at (x'). Write $\omega = \sum z_{(i)} dt_{(i)}$ where (t) is a set of local parameters at (x'). Say one of the coefficients $z_{(i)} = z$ is not in the local ring of (x'). Since this local ring is integrally closed by assumption, we can find a specialization $(x, 1/z) \rightarrow (x', 0)$. By the standard argument (Proposition 3 of Chapter VI, § 1) there is a subvariety W of codimension 1 containing (x') at which z has a pole. Thus z cannot be holomorphic at W.

In our preceding discussion, we have dealt with differential forms of a field K finitely generated over k, and viewed them as differential forms on an affine variety V, model of K. Of course, they can be viewed as differential forms on any abstract model of K. If $V = [V_\alpha, F_\alpha, T_\alpha^\beta]$ is such a model, and Q is a point of V, we shall say that Q is simple on V if for each V_α in which Q has a representative Q_α, the point Q_α is simple on V_α. We say V is non-singular if every point is simple. It will be shown in Chapter VIII that if one representative is simple, then so are the others, but we don't need to know this here. If $\mathfrak{o}$ is the local ring of a point of V in K, then we can say that a differential form ω is holomorphic at Q if it can be written $\omega = \sum z_{(i)} dy_{i_1} \wedge \ldots \wedge dy_{i_s}$ where the

$z_{(i)}$ and y_i are in $\mathfrak{o}$. This is compatible with our definition for affine varieties. We shall prove that if a differential form is holomorphic at every local ring belonging to a complete non-singular normal variety, then it is holomorphic at every local ring of a simple point of an affine model of K.

Let V be a complete abstract variety. Assume that every point of V is non-singular, and that V is normal (this latter assumption can be proved to be a consequence of the former). Then we shall say that a differential form is of *first kind* if it is holomorphic at every point of V. As a corollary to Theorem 4 we get the birational invariance of differential forms of first kind.

COROLLARY. *Let ω be a differential form of first kind on a complete non-singular normal variety. Let U be an affine normal variety birationally equivalent to V. Then ω is holomorphic at every simple point of U.*

Proof: Let P be a generic point of V over k. By hypothesis, the field $k(P)$ can be written $k(y)$, where (y) is a generic point of U over k. If there is a simple point of U where ω is not holomorphic, then there exists a prime divisor A of U at which ω is not holomorphic. The local ring of a generic point (y') of A in $k(y)$ is a valuation ring. If φ is a place belonging to this valuation ring, then φ induces a point on V because V is complete. We may now assume that V is affine. Let (x) be a generic point of V over k, and (x') the point induced by φ on V. The local ring of (x') in $k(x)$ is contained in the valuation ring above. If ω were holomorphic at (x') then it would be so *a fortiori* at A, a contradiction.

We shall attach a divisor to differential forms. The definitions are done locally, and so we may assume that V is affine and normal.

Let W be a prime divisor on V, and ω a differential form of the function field K of V, $K = k(x)$. Let (t) be local parameters at W (i.e. at a generic point (x') of W), and write

$$\omega = \sum z_{(i)} dt_{(i)}.$$

Define

$$\operatorname{ord}_W \omega = \min_{(i)} \operatorname{ord}_W (z_{(i)}).$$

We contend that $\operatorname{ord}_W \omega$ does not depend on the selection of local parameters (t) at W. If (u) is another set of parameters, and $\omega = \sum y_{(i)} du_{(i)}$ then we can write

$$du_i = \sum_{\mu} v_{i\mu} \, dt_{\mu}$$

where $v_{i\mu}$ are in the local ring of (x') in $k(x)$ (i.e. defined at W). Hence

$$du_{i_1} \wedge \ldots \wedge du_{i_s} = \sum G_{(i)} \, dt_{i_1} \wedge \ldots \wedge dt_{i_s},$$

where $G_{(i)}$ are in the local ring of (x') in $k(x)$. Hence the $z_{(i)}$ are linear combinations of the $y_{(i)}$ with coefficients in this local ring. This shows that

$$\operatorname{ord}_W z_{(i)} \geqq \min \operatorname{ord}_W y_{(i)}.$$

Our contention follows by symmetry.

Given any prime divisor W of V we can therefore speak of the order of zero or pole of a differential form at this prime divisor.

We shall prove a result analogous to the fact that a function has only a finite number of zeros and poles, namely that a differential form has only a finite number of them. This will be an immediate consequence of the following statement.

LEMMA. *Let $K = k(x)$ and let (t) be any separating transcendence base for $k(x)$ over k. Then (t) are local parameters for all but a finite number of prime divisors of V.*

Proof: Consider (t, x) as the generic point of a new variety U which is also a model of K, and consider a basis for the ideal of polynomials $f_\alpha(T, X)$ vanishing on (t, x). The Jacobian matrix $\| \partial f_\alpha / \partial X_\mu \|$ has rank $n - r$ at (x) according to our results on derivations. If we select n polynomials $f_1, \ldots, f_n$ among the f_α such that $\det (\partial f_i / \partial x_\mu) = z \neq 0$, then we may view z as a function on V, and we know that z has only a finite number of zeros and poles among the prime divisors of V. If W is not among this finite number, then (t) will be a set of local coordinates at W, as desired.

Any differential form ω can be written

$$\omega = \sum z_{(i)} \, dt_{(i)}$$

where (t) is a separating transcendence base. We can compute

$\mathrm{ord}_W \omega$ for any W except a finite number by using this representation, and we know that the $z_{(i)}$ have only a finite number of zeros and poles. This shows that ω also has only a finite number.

If V is complete and normal, then we can attach in an obvious way a divisor (ω) to ω, defined by

$$(\omega) = \sum n_j W_j \qquad\qquad n_j = \mathrm{ord}_{W_j} \omega.$$

Consider in particular a form of degree r. We can then write $\omega = z\, dt_1 \wedge \ldots \wedge dt_r$. Suppose (t) is a set of local coordinates at W, and suppose (u) is another set of local coordinates at W. Write $\omega = y\, du_1 \wedge \ldots \wedge du_r$. Then $du_1 \wedge \ldots \wedge du_r = J(u, t)dt_1 \wedge \ldots \wedge dt_r$, where J is the Jacobian, and $\mathrm{ord}_W J = 0$, so $\mathrm{ord}_W z = \mathrm{ord}_W y$. The divisors attached to differential forms of degree r are all linearly equivalent, because we have for any function y,

$$(y\omega) = (y) + (\omega).$$

This class of linearly equivalent divisors is called the *canonical class*.

We conclude this chapter by describing how a differential form holomorphic at a simple subvariety induces a form on this subvariety.

PROPOSITION 4. *Let $K = k(x)$, where (x) is the generic point of V over k. Let (x') be a simple point of V. Let (t) be a set of parameters at (x'), and let ω be a differential form of degree 1 of K which is holomorphic at (x'). Then ω can be written*

$$\omega = \sum z_i dt_i = \sum R_j(x)dx_j,$$

where z_i and $R_j(x)$ are in the local ring of (x') in $k(x)$. And if we denote by (t') and (z') the specializations of (t) and (z) over $(x) \to (x')$, then the two forms

$$\sum z_i' dt_i' \quad and \quad \sum R_j(x')dx_j'$$

of the field $K' = k(x)$ are equal.

Proof: Since each t_i is in the local ring of (x') in $k(x)$, dt_i can be written in terms of dx_j and elements of that local ring. By the definition of local parameters, we can find polynomials $f_j(T, X)$

such that $f_j(t, x) = 0$ but $D(t', x') \neq 0$, where $D(T, X) = \det(\partial f_j / \partial X_\mu)$. As in Proposition 3, we can find polynomials $G_{j\mu}(T, X)$ such that

$$(1) \qquad dx_j = \frac{\sum G_{j\mu}(x)dt_\mu}{D(t, x)},$$

and by considering the same set of linear equations as in Proposition 2 specialized over $(x) \to (x')$, we have

$$dx_j' = \frac{\sum G_{j\mu}(x')dt_\mu}{D(t', x')}.$$

On the other hand, if we substitute the expression (1) in $\omega = \sum R_j(x)dx_j$ and use the linear independence of the dt_i over K, we see that

$$z_i = \frac{\sum_j R_j(x)G_{jk}(x)}{D(t, x)}$$

and from this we get

$$z_i' = \frac{\sum_j R_j(x')G_{ji}(x')}{D(t', x')}.$$

The formula we wanted is an immediate consequence of the above expression.

Suppose now that ω is a differential form of arbitrary degree. It is a linear combination of Grassman products of forms of degree 1. Say $\omega = \sum z_{(i)}dt_{(i)}$, where (t) are local parameters at (x'), and $z_{(i)}$ are in the local ring of (x') in $k(x)$. It is clear from Proposition 4 that ω induces a form $\omega' = \sum z_{(i)}' dt_{(i)}'$ on the function field $k(x')$ of the subvariety U of V which is the locus of (x') over k.

CHAPTER VIII

Theory of Simple Points

We deal here with the local properties of simple points. The first section gives some abstract lemmas on commutative rings, which are needed afterwards. We take the absolute point of view, dealing with varieties over the universal domain.

1. Auxiliary results on commutative rings

In this section, except for Proposition 4, we allow our rings to have divisors of 0. A local ring is then a commutative ring with unit element, having a unique maximal ideal. The complement of that ideal in the ring is then the set of units of the ring.

PROPOSITION 1. *Let* $\mathfrak{o}$ *be a local ring,* $\mathfrak{m}$ *its maximal ideal, and* E *a finitely generated module over* $\mathfrak{o}$. *Suppose* $\mathfrak{m}E = E$. *Then* $E = 0$.

Proof: We prove this by induction on the number of generators of E. Let $x_1, \ldots, x_n$ be a set of generators for E, so $E = \mathfrak{o}x_1 + \ldots + \mathfrak{o}x_n$. Since $\mathfrak{m}E = E$, we can write $x_n = a_1 x_1 + \ldots + a_n x_n$ with $a_i \in \mathfrak{m}$. Thus $(1 - a_n)x_n = a_1 x_1 + \ldots + a_{n-1}x_{n-1}$. Since $1 - a_n$ is invertible in $\mathfrak{o}$, we see that E can be generated by $n - 1$ elements. This concludes the proof.

COROLLARY 1. *Let* E *be a finitely generated module over the local ring* $\mathfrak{o}$, *and let* F *be a submodule. If* $E = F + \mathfrak{m}E$, *then* $E = F$.

Proof: The factor module E/F is finitely generated, and satisfies the hypothesis of Proposition 1. It is therefore 0, so $E = F$.

COROLLARY 2. *If* $x_1, \ldots, x_n$ *is a set of generators for* $\mathfrak{m}$ *mod* $\mathfrak{m}^2$ *over* $\mathfrak{o}$, *then it is also a set of generators for* $\mathfrak{m}$ *over* $\mathfrak{o}$.

Proof: Let $E = \mathfrak{m}$, and $F = \mathfrak{o}x_1 + \ldots + \mathfrak{o}x_n$. Our hypothesis states that $E = F + \mathfrak{m}E$, and the preceding corollary shows that $E = F$, i.e. $\mathfrak{m} = \mathfrak{o}x_1 + \ldots + \mathfrak{o}x_n$, was as to be shown.

The next proposition can also be stated for modules, but we limit ourselves to ideals.

PROPOSITION 2. (ARTIN-REES). *Let R be a Noetherian ring, and $\mathfrak{a}$, $\mathfrak{b}$, $\mathfrak{c}$ three ideals. There exists an integer r such that for all $n \geq r$, we have $\mathfrak{a}^n\mathfrak{b} \cap \mathfrak{c} = \mathfrak{a}^{n-r}(\mathfrak{a}^r\mathfrak{b} \cap \mathfrak{c})$.*

Proof: Let $\mathfrak{a} = (a_1, \ldots, a_s)$ and consider the polynomial ring $R[X_1, \ldots, X_s]$ in s variables. Let $y \in \mathfrak{c} \cap \mathfrak{a}^n\mathfrak{b}$. We can write

$$y = f^{(n)}(a_1, \ldots, a_s),$$

where $f^{(n)}(X_1, \ldots, X_s)$ is a form of degree n with coefficients in $\mathfrak{b}$. With the pair (y, n) we have therefore associated some form $f^{(n)}$. Let F be the set of all such $f^{(n)}$ for all (y, n). It generates an ideal in $R[X_1, \ldots, X_s]$, which has a finite basis, say $f_1, \ldots, f_m$ of forms of F, of degrees $d_1, \ldots, d_m$. We can then write for each $f^{(n)}$:

$$f^{(n)} = \sum_{i=1}^{m} g_i^{(n-d_i)}(X)f_i(X),$$

where $g_i^{(n-d_i)}$ is a form of degree $n - d_i$ (or is equal to 0 if $n < d_i$). This shows that

$$y = f^{(n)}(a) = \sum_{i=1}^{m} g_i^{(n-d_i)}(a)f_i^{(d_i)}(a),$$

whence $y \in \sum (\mathfrak{a}^{d_i}\mathfrak{b} \cap \mathfrak{c})\mathfrak{a}^{n-d_i}$. If we let r be the largest d_i, and assume $n \leq r$, we have

$$(\mathfrak{a}^n\mathfrak{b} \cap \mathfrak{c}) \subset \sum_{\mu=1}^{r} (\mathfrak{a}^\mu\mathfrak{b} \cap \mathfrak{c})\mathfrak{a}^{n-\mu}$$

$$\subset \mathfrak{a}^{n-r}(\mathfrak{a}^r\mathfrak{b} \cap \mathfrak{c}) \subset (\mathfrak{a}^n\mathfrak{b} \cap \mathfrak{c}),$$

as was to be shown.

PROPOSITION 3. (KRULL'S THEOREM). *Let $\mathfrak{o}$ be a Noetherian local ring, and $\mathfrak{m}$ its maximal ideal. Then $\cap_{\mu=1}^{\infty} \mathfrak{m}^\mu = 0$.*

Proof: We take $\mathfrak{c} =$ unit ideal in the preceding proposition, and let $\mathfrak{b} = \cap \mathfrak{m}^\mu$. Then $\mathfrak{b} \cap \mathfrak{m}^n = \mathfrak{b}$ for all n, and by the proposition, it is also equal to $(\mathfrak{b} \cap \mathfrak{m}^r)\mathfrak{m}^{n-r} = \mathfrak{b}\mathfrak{m}^{n-r}$. Since $\mathfrak{m}^{n-r}$ is contained in $\mathfrak{m}$, we take $\mathfrak{b}$ to be the finite $\mathfrak{o}$-module in Proposition 1, and apply this proposition. We get $\mathfrak{b} = 0$, as desired.

PROPOSITION 4. *Let $\mathfrak{o}$ be a local ring without divisors of 0, $\mathfrak{m}$ its maximal ideal, and $(a_1, \ldots, a_m)$ a set of generators for $\mathfrak{m}$. Assume that if $f(X)$ is a form of degree μ in $\mathfrak{o}[X_1, \ldots, X_m]$, and if $f(a)$ lies in $\mathfrak{m}^{\mu+1}$, then all coefficients of f lie in $\mathfrak{m}$. Then $\mathfrak{o}$ is integrally closed.*

Proof: Let x and y be two elements of $\mathfrak{o}$, and suppose x/y is integral over $\mathfrak{o}$. Using Proposition 3 on the residue class ring $\mathfrak{o}$ mod (y), we see that if we prove that x is in the ideal $(y) + \mathfrak{m}^\mu$ for all μ, then we can conclude that x is in (y), i.e. $x = yz$ for some z in $\mathfrak{o}$, and x/y is in $\mathfrak{o}$. We shall now prove this by showing that if x is in $(y) + \mathfrak{m}^\mu$, then x is in $(y) + \mathfrak{m}^{\mu+1}$.

If $f(X)$ is a form in $\mathfrak{o}[X]$, we shall denote by $\bar{f}$ the form obtained by reducing all coefficients of f mod $\mathfrak{m}$, so that $\bar{f}$ has coefficients in $\mathfrak{o}/\mathfrak{m}$. Given an element z of $\mathfrak{o}$ which is in $\mathfrak{m}^\mu$ and not in $\mathfrak{m}^{\mu+1}$, there exists a form f of degree μ in $\mathfrak{o}[X]$ such that $z - f(a_1, \ldots, a_m)$ lies in $\mathfrak{m}^{\mu+1}$, and the reduced form $\bar{f}$ is unique (this being a consequence of our hypothesis). We denote the reduced form belonging to z by $\bar{f}_z$.

Suppose we have $x = yz + w$, with w in $\mathfrak{m}^\mu$. Dividing by y we see that w/y is integral over $\mathfrak{o}$, and hence there exists an integer n such that all powers of w/y can be written as linear combinations of $1, w/y, \ldots, (w/y)^{n-1}$ with coefficients in $\mathfrak{o}$. Hence there exists an element c of $\mathfrak{o}$, such that $c(w/y)^s$ is in $\mathfrak{o}$ for all positive integers s. In other words, $cw^s \,\epsilon\, y^s \mathfrak{o}$. From this it follows that for each s we can find a form $\bar{g}$ such that

$$\bar{f}_c \bar{f}_w^s = \bar{f}_y^s \, \bar{g},$$

i.e., $\bar{f}_y^s$ divides $\bar{f}_c \bar{f}_w^s$ for all s, and consequently $\bar{f}_y$ divides $\bar{f}_w$. We can write

$$\bar{h} \, \bar{f}_y = \bar{f}_w,$$

for some form $\bar{h}$. We let $h(X)$ be a representative of $\bar{h}$ in $\mathfrak{o}[X]$, and f_y, f_w be representatives of $\bar{f}_y$ and $\bar{f}_w$ in $\mathfrak{o}[X]$. Then we have

$$f_w(a) \equiv w \qquad (\mathrm{mod}\ \mathfrak{m}^{\mu+1}).$$
$$h(a)y \equiv f_w(a)\,(\mathrm{mod}\ \mathfrak{m}^{\mu+1}).$$

Taking into account our original relation $x = yz + w$, we see that $x \equiv y(z + h(a)) \bmod \mathfrak{m}^{\mu+1}$, and this concludes the proof.

REMARK: In view of Proposition 9 below, we shall see that the local ring of a simple point is integrally closed.

2. Definition of a simple point

Let V^r be a variety defined over a field k in affine n-space S^n. We let Ω be the universal domain, and $\mathfrak{P}_\Omega$ the prime ideal of V in $\Omega[X]$. We let $\mathfrak{P}_k$ be the prime ideal of V in $k[X]$.

If (x) is a point and $f \in \Omega[X]$, we denote as usual by $\partial f/\partial x_j$ the partial derivative $\partial f/\partial X_j$ evaluated at (x).

Consider the ranks of the following matrices at a point (x) of V:

$$\varrho_1 = \operatorname{rank} ||\partial f_\alpha/\partial x_i||,$$

where f_α ranges over all elements of $\mathfrak{P}_\Omega$.

$$\varrho_2 = \operatorname{rank} ||\partial f_\alpha/\partial x_i||,$$

where f_α ranges over all elements of $\mathfrak{P}_k$.

$$\varrho_3 = \operatorname{rank} ||\partial f_j/\partial x_i||,$$

where f_j ranges over a finite basis of $\mathfrak{P}_k$.

Then it is clear that $\varrho_1 \geq \varrho_2 \geq \varrho_3$, and in fact we now show that these three ranks must be equal. A polynomial f_α in $\mathfrak{P}_\Omega$ can be written

$$f_\alpha = g_1 f_1 + \ldots + g_m f_m$$

with f_j in $\mathfrak{P}_k$. Taking the partials and noting that $f_j(x) = 0$ because (x) is assumed to be in V, we see that the contribution of f_α to the first matrix depends linearly on the contribution of the f_j and hence cannot increase the rank. Hence $\varrho_1 = \varrho_2$. The equality $\varrho_2 = \varrho_3$ is proved in the same way.

If (x) is a generic point of V/k, then we conclude from Chapter VII, § 1 that $\varrho_3 = n - r$. Furthermore, if all elements of a matrix are specialized simultaneously over a field k, then the rank of the specialized matrix obviously cannot increase. This implies that for an arbitrary point (x) of V, the rank of the above matrix is $\leq n - r$. We define the point (x) to be *simple on* V if the rank is exactly $n - r$.

The above discussion shows two things: First, that our definition is geometric, i.e. does not depend on the field of definition selected for V, and second that a generic point of V is always simple on V. A point which is not simple is said to be *singular*.

PROPOSITION 5. *Let V be a variety defined over a field k. Then the set of singular points on V is a proper k-closed subset of V.*

Proof: Let f_j be a basis for $\mathfrak{P}_k$ and let $D_\nu(X)$ be all the determinants of order $n - r$ belonging to the matrix $||\partial f_j/\partial X_i||$. The set of singular points is the intersection of V with the algebraic set of zeros $D_\nu(X) = 0$, and is therefore k-closed. In addition we have already seen that it cannot be all of V.

For want of a better place, we state here the proposition dealing with a product of points.

PROPOSITION 6. *Let V^r and $W^{r'}$ be two varieties in S^n and $S^{n'}$ respectively, and let P be a point of V and Q be a point of W. Then (P, Q) is simple on $V \times W$ if and only if P is simple on V and Q is simple on W.*

Proof: Let V and W be defined over k. If $f_\alpha(X)$ is a basis for the ideal of V in $k[X]$ and $g_\beta(Y)$ is a basis for the ideal of W in $k[Y]$ then the two combined form a basis for $V \times W$ in $k[X, Y]$ according to Proposition 1 of Chapter IV, § 1. If we now look at the Jacobian matrix at the point (P, Q) on $V \times W$, we see that it can be of rank $n + n' - r - r'$ if and only if P is simple on V and Q is simple on W.

Our definition of simplicity has had the advantage of showing immediately that the set of singular points is an algebraic set and is independent of fields of definition. It has the disadvantage that it is not immediately obvious that the property of a point to be simple is invariant under biholomorphic transformations. This will be proved in the next section.

3. Existence of local uniformizing parameters

In this entire section, V is a variety of dimension r in S^n, defined over k. We always denote by (x) a generic point of V/k and by (x') an arbitrary point of V. We let s be the dimension of (x') over k. We recall that the local ring of (x') in $k(x)$ consists of all quotients $f(x)/g(x)$ where f, g are polynomials with coefficients in k, and $g(x') \neq 0$. This local ring will be denoted by $\mathfrak{o}$ and its maximal ideal by $\mathfrak{m}$. The locus of (x') over k is a k-subvariety of V which will be denoted by W.

PROPOSITION 7. *The ideal $\mathfrak{m}$ has at least $r - s$ generators. In other words, if $\mathfrak{m} = (y_1, \ldots, y_m)$ then $s \geq r - m$.*

Proof: Each generator is a quotient of two polynomials, and the denominator does not vanish at (x'), hence is a unit in $\mathfrak{o}$. Without loss of generality we may therefore assume that $y_j = f_j(x)$ where $f_j(X) \in k[X]$, and f_j vanishes on the point (x'), i.e. $f_j(x') = 0$. We intersect V with the hypersurfaces $f_1 = 0, \ldots, f_m = 0$. Then W is contained in this intersection. Let W^* be a k-component of this intersection containing W. By the dimension theorem, $\dim W^* \geq r - m$. To prove that $W^* = W$, it will suffice to prove that $W^* \subset W$. Let $\mathfrak{p}^*$ and $\mathfrak{p}$ be the prime ideals of W^* and W respectively in the coordinate ring $R = k[x]$. Then $R \supset \mathfrak{p} \supset \mathfrak{p}^*$. We must show that $\mathfrak{p} \subset \mathfrak{p}^*$. Let $g \in \mathfrak{p}$. Then $g \in \mathfrak{o}$ and we can write

$$g = (h_1/g_1)f_1 + \ldots + (h_m/g_m)f_m,$$

where h_j, g_j lie in R but $g_j \notin \mathfrak{p}$, hence $g_j \notin \mathfrak{p}^*$. Let G be the product of the g_j. Then G is not in $\mathfrak{p}^*$ either, but Gg is in $\mathfrak{p}^*$. Hence g is in $\mathfrak{p}^*$ and this concludes the proof.

If the maximal ideal of the local ring of (x') has exactly $r - s$ generators, then we shall say that $\mathfrak{o}$ is *regular*. It will be proved later that the local ring of a simple point is regular, and that if in addition $k(x')$ is separably generated over k then the converse also holds. This will give us the desired characterization of simple points.

The generators y_j in the preceding proposition could be written $y_j = f_j(x)/g_j(x)$ where $g_j(x') \neq 0$. If W^* is a k-subvariety of V containing W, and (x^*) a generic point of W^* over k, then *a fortiori* we must have $g_j(x^*) \neq 0$, i.e. the y_j are defined at W^*. An argument similar to the one used to prove the preceding proposition yields the following result:

PROPOSITION 8. *As before, let* $(y_1, \ldots, y_m)$ *be a set of generators for* $\mathfrak{m}$. *If* W^* *is a* k-*subvariety of* W *containing* W, *and if* $y_1, \ldots, y_m$ *vanish on* W^*, *then* $W^* = W$.

In particular, if $m = r - s$ in the above proposition, then one interprets this result as stating that W is locally a complete intersection of V and the hypersurfaces $f_j = 0$. In view of Theorem 1 below, one sees that a simple subvariety of a variety V is always locally a complete intersection.

We shall now prove one of the main theorems of this section, namely

THEOREM 1. *The local ring of a simple point is regular.*

Our first step in the proof will be to settle the special case where we deal with affine space itself. In other words, we have the following lemma.

LEMMA. *Let* $V = S^n$ *be affine space itself. Let* (x') *be a point in* S^n, *of dimension* s *over* k, *and let* (x) *be* n *independent variables over* k, *so* (x) *is a generic point of* S^n *over* k. *Let* $\mathfrak{o}$ *be the local ring of* (x') *in* $k(x)$, *and* $\mathfrak{m}$ *its maximal ideal. Then* $\mathfrak{m}$ *has* $n - s$ *generators.*

Proof: Say $x'_1, \ldots, x'_s$ is a transcendence basis for $k(x')$ over k. Then after a suitable isomorphism of $k(x')$ over k, we may assume that $x'_i = x_i$ for $i = 1, \ldots, s$. Let $k_1 = k(x_1, \ldots, x_s)$. Then $k(x')$ is algebraic over k_1, and $k_1(x_{s+1}, \ldots, x_n)$ is purely transcendental over k_i. It is easily seen that the local ring of $(x'_{s+1}, \ldots, x'_n)$ in $k_1(x_{s+1}, \ldots, x_n)$ is equal to $\mathfrak{o}$. We are therefore brought to proving our lemma in case our point (x') is algebraic over the constant field.

We can therefore work with S^n, and a point (x') such that $k(x')$ is algebraic over k. Let $K_i = k[x'_1, \ldots, x'_i]$, $i = 1, \ldots, n$. It is not only a ring, but also a field because it is algebraic over k. Let $f_i(X_1, \ldots, X_i) \in k[X_1, \ldots, X_i]$ be polynomials such that $f_i(x'_i, \ldots, x'_{i-1}, X_i)$ is the irreducible polynomial for x'_i over K_i, and such that the leading coefficient in X_i is equal to 1. Then $f_i(X)$ vanishes on (x'), and hence lies in the ideal $\mathfrak{p}$ of (x') in $k[x]$. It will suffice to prove that the f_i generate $\mathfrak{p}$. Suppose g is in $\mathfrak{p}$, so $g(x') = 0$. Since f_n has leading coefficient 1 in X_n, we can write

$$g(X) = f_n(X)Q_n(X) + R_n(X),$$

where Q_n and R_n lie in $k[X_1, \ldots, X_n]$, and the degree of R_n relative to X_n is strictly less than that of f_n. We have $R_n(x') = 0$. Since $f_n(x'_1, \ldots, x'_{n-1}, X_n)$ is the irreducible polynomial of x_n over K_{n-1}, it follows that $R_n(x'_1, \ldots, x'_{n-1}, X_n)$ is identically zero. Consequently we can write

$$R_n(X) = \sum P_{nj}(X_1, \ldots, X_{n-1})X_n^j$$

where P_{nj} are in $k[X_1, \ldots, X_{n-1}]$ and $P_{nj}(x') = 0$. The lemma now follows by induction, working in the polynomial rings $k[X_1, \ldots, X_i]$ $(i = n - 1, \ldots, 1)$.

Having settled the case where $V = S^n$, we shall now base the rest of the investigations of this section on a comparison between the local ring $\mathfrak{o}$ of a point of V, and its local ring when the point is viewed as a point of S^n. We therefore fix the terminology and notation which will be used in the rest of this section.

We denote by $(X) = (X_1, \ldots, X_n)$ the variables of S^n, and if (x') is as before a point of V, then we let $\mathfrak{O}$ be the local ring of (x') in $k(X)$, and $\mathfrak{M}$ its maximal ideal. In $k[X]$ we have two prime ideals $\mathfrak{p} \supset \mathfrak{q}$ determined by (x') and (x) respectively. The reader will verify immediately that $\mathfrak{q}\mathfrak{O}$ is a prime ideal $\mathfrak{N}$, and that $\mathfrak{O}/\mathfrak{N}$ is canonically isomorphic to the local ring $\mathfrak{o}$ of (x') in $k(x)$, under the mapping

$$\frac{f(X)}{g(X)} \to \frac{f(x)}{g(x)}.$$

The image of $\mathfrak{M}$ is precisely the maximal ideal $\mathfrak{m}$ of $\mathfrak{o}$.

If D is a derivation of $\mathfrak{O}$ into itself, then D induces in a natural way a linear form of the vector space $\mathfrak{M}/\mathfrak{M}^2$ over $\mathfrak{O}/\mathfrak{M}$. Namely, D certainly maps $\mathfrak{M}$ into $\mathfrak{O}$, and the mapping is linear. Let $\tau : \mathfrak{O} \to \mathfrak{O}/\mathfrak{M}$ be the canonical homomorphism. Then the composite map $\tau \circ D$ vanishes on $\mathfrak{M}^2$, (this is an immediate consequence of the definition of a derivation) and hence induces a $\mathfrak{O}/\mathfrak{M}$-linear form of $\mathfrak{M}/\mathfrak{M}^2$ over $\mathfrak{O}/\mathfrak{M}$. In the applications, D is one of the derivations $\partial/\partial X_i$. These map $\mathfrak{O}$ into $\mathfrak{O}$, as one deduces from the rule $D(u/v) = (vDu - uDv)/v^2$.

These induced linear forms give us a mechanism which transforms a problem concerning the non-vanishing of a Jacobian into a linear problem concerning the vector space $\mathfrak{M}/\mathfrak{M}^2$ over $\mathfrak{O}/\mathfrak{M}$, and *vice versa*. Indeed, let $\partial/\partial X_\alpha$ range over a subset of the derivations $\partial/\partial X_i$, and let $f_\beta(X)$ be a finite number of polynomials in $\mathfrak{M}$. Let λ_α be the linear form induced by $\partial/\partial X_\alpha$, and c_β the residue class of $f_\beta(X)$ mod $\mathfrak{M}^2$. Then it is immediate from the definitions that

$$\text{rank } ||\partial f_\beta/\partial x'_\alpha|| = \text{rank } ||\lambda_\alpha(c_\beta)||.$$

With the preceding notations, we consider the following inclusion:

$$\mathfrak{M} \supset \mathfrak{N} + \mathfrak{M}^2 \supset \mathfrak{M}^2.$$

Then $\mathfrak{M}/(\mathfrak{N} + \mathfrak{M}^2)$ is a vector space over $\mathfrak{O}/\mathfrak{M}$. If we can prove that it has at most $r - s$ generators, then $\mathfrak{M}$ mod $(\mathfrak{N} + \mathfrak{M}^2)$ has at most $r - s$ generators over $\mathfrak{O}$, and all we need to do to prove Theorem 1 is to read this statement mod $\mathfrak{N}$. We shall prove this by showing that $\mathfrak{N} + \mathfrak{M}^2/\mathfrak{M}^2$ is of dimension at least $n - r$ over $\mathfrak{O}/\mathfrak{M}$, and by using the lemma, which shows that $\mathfrak{M}/\mathfrak{M}^2$ has dimension $n - s$.

According to the definition of a simple point, and after a suitable renumbering of the variables, we can find $n - r$ polynomials $f_j(X)(j = r + 1, \ldots, n)$ vanishing on V (and consequently in $\mathfrak{N}$) and such that the determinant

$$\det (\partial f_j/\partial X_\nu) \qquad j, \nu = r + 1, \ldots, n$$

does not vanish at (x'). Let D_j be the derivations $\partial/\partial X_j$. They map $\mathfrak{O}$ into $\mathfrak{O}$, and induce linear forms λ_j of $\mathfrak{M}/\mathfrak{M}^2$ over $\mathfrak{O}/\mathfrak{M}$.

Let c_j denote the residue classes of f_j mod $\mathfrak{M}^2$. Then the non-vanishing of the above determinant implies that

$$\det (\lambda_\nu(c_j))$$

is not 0. This means that the c_j are linearly independent over $\mathfrak{O}/\mathfrak{M}$. Since the f_j are in $\mathfrak{N}$, this shows that the dimension of $\mathfrak{N} + \mathfrak{M}^2/\mathfrak{M}^2$ over $\mathfrak{O}/\mathfrak{M}$ is at least $n - r$ and concludes the proof of Theorem 1.

Our next task is to prove the invariance of simple points under biholomorphic transformations. In order to do this, we remark that we can work over any field by Chapter IV, § 3. Hence we may assume without loss of generality that the point (x') is rational over k.

In that case, we have

THEOREM 2. *Assume that (x') is rational over k. If the local ring $\mathfrak{o}$ of (x') in $k(x)$ is regular, then (x') is simple on V.*

Proof: The proof consists in showing that all the steps in the proof of Theorem 1 are reversible, because we have taken (x') rational over k. We carry these steps out in detail.

Our hypothesis that $\mathfrak{o}$ is regular means that m has $r - s$ generators, or, put another way, that $\mathfrak{M}/\mathfrak{N} \bmod (\mathfrak{N} + \mathfrak{M}^2)/\mathfrak{N}$ has $r - s$ generators over $\mathfrak{O}/\mathfrak{N}$. From this one sees that $\mathfrak{M} \bmod (\mathfrak{N} + \mathfrak{M}^2)$ has $r - s$ generators over $\mathfrak{O}$, and since $\mathfrak{M}/\mathfrak{M}^2$ has $n - s$ generators over $\mathfrak{O}/\mathfrak{M}$, it follows that $\mathfrak{N} + \mathfrak{M}^2/\mathfrak{M}^2$ is a vector space of dimension $n - r$ over $\mathfrak{O}/\mathfrak{M}$. We can therefore find $n - r$ linearly independent elements $c_j (j = r + 1, \ldots, n)$ of $\mathfrak{N} + \mathfrak{M}^2/\mathfrak{M}^2$ which are represented in $\mathfrak{N}$ by polynomials $f_j(X)$.

In the present case where (x') is rational over k, we have $s = 0$. Furthermore, the n derivations $\partial/\partial X_i$ in $k(X)$ induce n linearly independent linear forms $\lambda_i (i = 1, \ldots, n)$ of $\mathfrak{M}/\mathfrak{M}^2$ over $\mathfrak{O}/\mathfrak{M}$. It is to prove this that we make use of our assumption that (x') is rational over k, and hence $\mathfrak{O}/\mathfrak{M}$ is canonically isomorphic to k. If we have a linear relation $\sum a_i \lambda_i = 0$ with $a_i \in k$, then the derivation $\sum a_i (\partial/\partial X_i)$ maps $\mathfrak{M}$ into $\mathfrak{M}$. Since the elements $X_i - x_i$ form a basis for $\mathfrak{M}$ over $\mathfrak{O}$, this can happen only if all $a_i = 0$. Hence our induced linear forms are linearly independent over k.

Consequently we can select $n - r$ among the λ_i, say for $i = r + 1, \ldots, n$, such that $\det (\lambda_\nu(c_j)) \neq 0$. This implies that $\det (\partial f_j/\partial x'_\nu) \neq 0$. Since the f_j were selected in $\mathfrak{N}$, they vanish on V, and this proves that (x') is simple on V.

In view of the remark made before the theorem, we get

COROLLARY. *If U and V are biholomorphic at points P and Q, then P is simple on U if and only if Q is simple on V.*

The following theorem gives us the existence of the parameters used in the theory of differential forms.

THEOREM 3. *Let (x') be a simple point of V, and assume that $k(x')$ is separably generated over k. Let $(t_1, \ldots, t_r) = (t)$ be a set of elements of $\mathfrak{o}$, and (t') the specialization of (t) over $(x) \to (x')$. Assume that (t) has the following properties:*

1. *$(t'_1, \ldots, t'_s)$ form a separating transcendence base of $k(x')$ over k, and*

2. *$t_{s+1}, \ldots, t_r$ form a basis of $\mathfrak{m}$.*
Then there exist polynomials $f_i(T_1, \ldots, T_r, X_1, \ldots, X_n) = f_i(T, X)$ $(i = 1, \ldots, n)$ such that

$$f_i(t, x) = 0 \quad but \quad \det (\partial f_i/\partial x'_\nu) \neq 0.$$

Proof: Without loss of generality we may assume that the quantities (t) are the first r coordinates of (x), i.e. that we have $x_1 = t_1, \ldots, x_r = t_r$. Indeed, the locus of (t, x) over k is biholomorphic to V at (t', x') and (x'). Thus we may replace V by this locus. Using this added assumption, we must prove the existence of polynomials $f_j(X) (j = r + 1, \ldots, n)$ such that $f_j(x) = 0$ (i.e., f_j vanish on V) but $\det (\partial f_j/\partial x'_\nu) \neq 0$ $(j, \nu = r + 1, \ldots, n)$.

Since $(x_{s+1}, \ldots, x_r)$ generate $\mathfrak{m}$, it follows that $(X_{s+1}, \ldots, X_r)$ lie in $\mathfrak{M}$, and in fact are linearly independent mod $(\mathfrak{N} + \mathfrak{M}^2)$ over $\mathfrak{O}/\mathfrak{M}$. Consequently $X_{s+1}, \ldots, X_r$ can be completed to a basis of $\mathfrak{M}/\mathfrak{M}^2$ over $\mathfrak{O}/\mathfrak{M}$ by means of polynomials $f_j(X) (j = r + 1, \ldots, n)$ which lie in $\mathfrak{N}$, i.e. vanish on V. We shall see that these will achieve what we want. We let $f_\alpha(X) (\alpha = s + 1, \ldots, n)$ be the $n - s$ polynomials $X_{s+1}, \ldots, X_r, f_{r+1}, \ldots, f_n$.

Since $k(x')$ is separably algebraic over $k(x'_1, \ldots, x'_s)$ by hypothesis, there exist polynomials $g_j(X_1, \ldots, X_s, X_j)$ such that $g_j(x') = 0$ but $\partial g_j/\partial x'_j \neq 0$. This implies that the $n - r$ derivations $\partial/\partial X_j (j = r + 1, \ldots, n)$ induce $n - r$ linearly independent forms λ_j of $\mathfrak{M}/\mathfrak{M}^2$ over $\mathfrak{O}/\mathfrak{M}$.

Combining the remarks of the last two paragraphs, we conclude that the rank of the matrix

$$||\partial f_\alpha / \partial x'_j||$$

$$\alpha = s + 1, \ldots, n$$
$$j = r + 1, \ldots, n$$

is at least $n - r$. Since $f_\alpha = X_\alpha$ for $\alpha = s + 1, \ldots, r$ we see that the Jacobian det $(\partial f_j / \partial X_\nu)(j, \nu = r + 1, \ldots, n)$ cannot vanish at (x'). Recalling that the f_j were selected in $\mathfrak{N}$ and thus vanish on V, we see that our theorem is proved.

We shall give a converse of Theorem 3 only in the case where (x') is a rational point. In that case, we get a geometric characterization for a set of generators of $\mathfrak{m}$ (i.e. a characterization independent of the ground field).

THEOREM 4. *If (x') is a point of V rational over k, and if there exist elements $t_1, \ldots, t_r$ in $\mathfrak{m}$ and polynomials $f_i(T, X)(i = 1, \ldots, n)$ such that*

$$f_i(t, x) = 0 \qquad but \qquad \det (\partial f_i / \partial x'_\nu) \neq 0,$$

then (x') is simple on V and $(t_1, \ldots, t_r)$ generate $\mathfrak{m}$.

Proof: We observe again that we may assume that the $t_1, \ldots, t_r$ are the first r coordinates of (x), because the property of a point to be simple is biregularly invariant. The hypothesis of the theorem then implies that the residue classes mod $\mathfrak{m}^2$ of these first r coordinates are linearly independent over $\mathfrak{o}/\mathfrak{m}$. By definition, they also imply that the point is simple, and hence these first r coordinates must be a basis of $\mathfrak{m}$, as was to be shown.

4. The expansion in power series

The notation remains the same as in the preceding section.

The possibility of expanding a function into a power series at a simple point will follow from the regularity of the local ring, and from the following proposition.

PROPOSITION 9. *Let (x') be a simple point of V, again of dimension s over k. Let $t_1, \ldots, t_m (m = r - s)$ be a set of generators for the maximal ideal $\mathfrak{m}$ of the local ring $\mathfrak{o}$. If $f(T)$ is a form of degree ν with coefficients in $\mathfrak{o}$ such that $f(t) \equiv 0 \bmod \mathfrak{m}^{\nu+1}$ then all coefficients of f lie in $\mathfrak{m}$.*

Proof: We keep the notation of § 1, and if a is an element of $\mathfrak{o}$

then $\bar{a}$ denotes its residue class mod $\mathfrak{m}$. We denote by $\bar{f}$ the form obtained from f by reducing all its coefficients mod $\mathfrak{m}$.

We shall first prove our proposition in the case where $\mathfrak{o}/\mathfrak{m}$ is an infinite field. In that case, there exist elements $(\bar{u}_{ij})(i, j = 1, \ldots, m$ in $\mathfrak{o}/\mathfrak{m}$ such that if (u_{ij}) is a set of representatives in $\mathfrak{o}$ and if we put

$$z_i = \sum u_{ij} t_j$$

then z_m satisfies an equation

$$z_m^d + a_{m-1} z_m^{d-1} + \ldots + a_0 = 0$$

with coefficients a_μ in the ideal $(z_1, \ldots, z_{m-1})$, and such that the linear transformation $(\bar{u}_{ij})$ is non-singular. From this it follows that $(z_1, \ldots, z_m)$ is again a basis of $\mathfrak{m}/\mathfrak{m}^2$ over $\mathfrak{o}/\mathfrak{m}$. Without loss of generality we may assume that $z_1, \ldots, z_m$ are among the coordinates of the generic point (x) of V/k.

Intersect V with the hypersurfaces $z_1 = 0, \ldots, z_{m-1} = 0$, and let W^* be a component of this intersection containing W. Then the coefficients a_μ being defined at W are also defined at W^*, and so is z_m. In view of the above equation, we see that z_m must vanish on W^*, and hence by Proposition 8 we must have $W^* = W$. This contradicts the dimension theorem, according to which dim $W^* \geq r - (m - 1) > s$.

In case the residue class field $\mathfrak{o}/\mathfrak{m}$ is finite, we use a standard device of adjoining a transcendental element u to the field $k(x, x')$. Put $k_u = k(u)$. The reader will immediately verify that if $\mathfrak{o}_u$ and $\mathfrak{m}_u$ denote the local ring of (x') in $k_u(x)$ and its maximal ideal, then we have

$$\mathfrak{m}_u = \mathfrak{m}\mathfrak{o}_u \quad \text{and} \quad \mathfrak{m} = \mathfrak{m}_u \cap \mathfrak{o}.$$

From this it follows that if our proposition holds with respect to the field k_u then it holds with respect to the field k. This concludes the proof of our proposition.

Assume that our simple point is a rational point, and let $t_1, \ldots, t_r$ be generators of $\mathfrak{m}$. Let w be an arbitrary element of the local ring $\mathfrak{o}$. There exists a constant c_0 such that $w \equiv c_0$ mod $\mathfrak{m}$. By the preceding proposition, there exists a unique linear form $f_1(T)$ with coefficients in k such that $w - c_0 \equiv f_1(t) \, (\text{mod } \mathfrak{m}^2)$.

Again by the proposition, there exists a unique form $f_2(T)$ of degree 2 with coefficients in k such that

$$w \equiv c_0 + f_1(t) + f_2(t) \pmod{\mathfrak{m}^3}.$$

Proceeding in this fashion, we get a mapping of w onto a well defined formal power series in the power series ring $k[[t]]$, which clearly gives a homomorphism of $\mathfrak{o}$ into this power series ring.

If an element w maps on 0 under this homomorphism, then w must be in the intersection $\cap_{\nu=1}^\infty \mathfrak{m}^\nu$, and hence must be equal to 0 by Proposition 3 of § 1. Our homomorphism is therefore an isomorphism, and gives us the desired power series expansion.

5. Dimension theorem for simple components

We wish to generalize here the dimension theorem proved in Chapter II, § 7.

THEOREM 5.   *Let A^r, B^s be two subvarieties of a variety U^n, and let P be a point in $A \cap B$, simple on U. Then there exists a component C of $A \cap B$, containing P, and of dimension $\geq r + s - n$.*

From this, we shall obtain immediately the following result:

COROLLARY.   *If C is a component of $A \cap B$ which is simple on U, then its dimension is $\geq r + s - n$.*

Proof: We take a simple point P of C which does not lie in any other component of $A \cap B$, for instance a generic point of C over a common field of definition for all the components of $A \cap B$, and we apply the theorem.

The proof of the theorem is based on the simple geometric idea of projecting locally A, B, and U on the tangent n-space of U at P, and of using the known result for affine n-space.

To begin with, we have a theorem which will give us a criterion for such a projection to occur locally in a $1 - 1$ manner.

THEOREM 6.   *Let (x, y) and (x, z) be two points in $S^n \times S^m$ and let $(x, y) \to (x', y')$ be a specialization over a field k. Assume that we have polynomials $f_i(X, Y)$ in $k[X, Y]$ $(i = 1, \ldots, m)$ vanishing on both (x, y) and (x, z). Assume in addition that if we put $D(X, Y)$*

$= \det (\partial f_i / \partial Y_j)$ *then* $D(x', y') \neq 0$. *Then we cannot have a specialization*

$$(x, y, z) \to (x', y', y') \quad \text{over} \quad k$$

unless $(y) = (z)$.

Proof: Using vector notation, let $(w) = (z) - (y)$. We can write

$$0 = f_i(x, y + w) = f_i(x, y) + \sum \frac{\partial f_i}{\partial y_\nu} w_\nu + \sum \frac{\partial^2 f_i}{\partial y_\nu \, \partial y_\mu} w_\nu \, w_\mu + \cdots$$

By assumption, $f_i(x, y) = 0$. All the terms in the above sum after the first term (linear in the w's) are homogeneous of degree ≥ 2 in the w's. Take a place φ extending the specialization $(x, y, z) \to (x', y', y')$, and say w_1 has maximal value under this place. Divide the above equation by w_1 and apply the place. We have $\varphi(w) = (0)$, and all $\varphi(w_\nu / w_1)$ are finite. Hence all the terms of degree ≥ 2 go to 0, and we must have

$$0 = \sum (\partial f_i / \partial y'_\nu) c_\nu,$$

where $c_\nu = \varphi(w_\nu / w_1)$. This contradicts the hypothesis concerning the non-vanishing of the determinant, and proves our theorem.

We now describe the generic projection of our variety U on S^n. We let S^N be the affine space in which U lies.

Let k be a common field of definition for A, B, U, and assume also that P is rational over k. Without loss of generality we may then assume that P is the origin on S^N, $P = (0, \ldots, 0)$.

Let (x) be a generic point of U/k and let $(u_{i\nu})$ be independent variables over $k(x)$, with $i, \nu = 1, \ldots, N$. Put

$$\bar{x}_i = \sum u_{i\nu} x_\nu \qquad i = 1, \ldots, N.$$

Let K be the algebraic closure of $k(u)$. Then the locus of $(\bar{x})$ over K is a variety U_0 obtained from U by what is known as generic linear transformation. It is obviously biholomorphic to U over K, and the origin P on U corresponds to the origin P_0 on U_0. The varieties A and B have corresponding subvarieties A_0 and B_0 of U_0.

Since P is simple on U, P_0 must be simple on U_0 (either by a direct argument from the definition of simplicity, or by using the biholomorphic invariance). After a suitable renumbering of the coordinates, there exists polynomials $f_j(X) \in K[X]$ $(j = n + 1, \ldots, N)$ vanishing on $(\bar{x})$ (i.e. on U_0) and such that the the the determinant $\det (\partial f_j/\partial X_\alpha)$ for $j, \alpha = n + 1, \ldots, N$ does not vanish at P_0.

We break up our coordinates into two groups, namely

$$y_i = \sum u_{i\nu} x_\nu \qquad\qquad i = 1, \ldots, n$$
$$z_j = \sum u_{j\nu} x_\nu \qquad\quad j = n + 1, \ldots, N$$

and split S^N accordingly: $S^N = S^n \times S^{N-n}$. Then the following properties hold:

1.   The coordinates z_j are integral over $K[y]$ and in particular the projection of U_0 on the first factor S^n is S^n.

This is well known from the usual way of proving the Noether normalization theorem.

2.   Let A_0' and B_0' be the projections of A_0 and B_0 on S^n. Then $\dim A_0' = \dim A_0$ and $\dim B_0' = \dim B_0$. Furthermore, to any point (y') on A_0' there corresponds at least one point (y', z') on A_0 projecting on (y').

This is an immediate consequence of Property 1.

3.   If (y', z') and (y', w') are two points of U_0 and if $(y', z', w') \to (0, 0, 0)$ is a specialization over K, then $(z') = (w')$.

This is simply a direct application of the criterion given in Theorem 6, and of our hypothesis concerning the non-vanishing of a Jacobian.

4.   Assume without loss of generality that neither A nor B are equal to U. If $(0, \xi)$ in $S^n \times S^{N-n}$ is a point of A_0 or of B_0 then $(\xi) = (0)$. In other words, the only point of A_0 or B_0 in S^N lying above the point (0) in S^n is $(0, 0) = P_0$.

This is proved as follows. Let (y', z') be a generic point of A_0 over K, and let (x') be the corresponding generic point of A over K. Then

$$y_i' = \sum u_{i\nu} x_\nu' \qquad\qquad i = 1, \ldots, n$$

and (x') is integral over $K[y']$. The point $(0, \xi)$ is a specialization of (y', z') over K, and this specialization can be extended to a specialization $(x') \to (\xi)$. It will suffice to prove that $(\xi) = (0)$. We can write

$$0 = u_{11}\xi_1 + \ldots + u_{1N}\,\xi_N$$
$$0 = u_{n1}\xi_1 + \ldots + u_{nN}\,\xi_N.$$

If, say, ξ_1 is not 0, then we divide by ξ_1 throughout. This shows that $u_{11}, \ldots, u_{n1}$ can be expressed rationally over k in terms of the other u's above, and in terms of (ξ). However (ξ) is a point in A, and hence its transcendence degree over k is $< n$ because we have assumed $A \neq U$. This gives a contradiction and proves our fourth property.

Let D' be a component of $A'_0 \cap B'_0$ containing (0) on S^n. Then $\dim D' \geq r + s - n$ by the ordinary dimension theorem. We shall prove that there exists a variety D in $A_0 \cap B_0$ containing P_0 whose projection on S^n is D'. Since $\dim D \geq \dim D'$ (in fact the two dimensions are equal) and since D is contained in a component whose dimension is *a fortiori* $\geq r + s - n$, our theorem will be proved.

Let (y') be a generic point of D' over K. Then (y') is a point of A'_0 and of B'_0. According to Property 2, we can find a point (y', z') on A_0 and a point (y', w') on B_0 projecting on (y'). We contend that $(z') = (w')$. Indeed, we know that $(y') \to (0)$ is a specialization over K. Extend this to a specialization $(y', z', w') \to (0, \xi, \omega)$. Then the point $(0, \xi)$ lies in A_0. By Property 4 we must have $(\xi) = (0)$. Similarly $(\omega) = (0)$. By Property 3 we must have $(z') = (w')$.

Now (y', z') is the generic point of some variety D over K, and D is contained in $A_0 \cap B_0$ because $(y', z') \in A_0 \cap B_0$. Its projection on S^n is D', and $P_0 \in D$. This is just what we wanted, and concludes the proof of our generalized dimension theorem.

6. The generic hyperplane section

For this entire section, we shall adopt the following notation. V^r is an affine variety in S^n, defined over a field k and of dimension $r \geq 1$. We let $u_1, \ldots, u_n$ be independent quantities over k, and let (x) be a generic point of V over $k(u_1, \ldots, u_n)$. Let

$$u_{n+1} = u_1 x_1 + \ldots + u_n x_n.$$

We let $k_u = k(u) = k(u_1, \ldots, u_{n+1})$, and denote by W_u the locus of (x) over k_u. It is a k_u-variety.

We first observe that u_{n+1} is transcendental over $k(u_1, \ldots, u_n)$. Indeed, u_{n+1} is contained in $k(u_1, \ldots, u_n, x)$ which is a regular extension of $k(u_1, \ldots, u_n)$. If it were not transcendental over $k(u_1, \ldots, u_n)$ it would lie in that field. We can specialize (x) to two distinct points (x') and (x'') algebraic over k, while keeping $(u_1, \ldots, u_n)$ fixed, and hence u_{n+1} fixed. (We view $(x, u_1, \ldots, u_n)$ as a generic point of a product.) Subtracting the linear equations expressing u_{n+1} in terms of $(u_1, \ldots, u_n)$, (x'), and (x''), would give a linear relation among $(u_1, \ldots, u_n)$ with coefficients algebraic over k, thereby contradicting their algebraic dependence over k.

THEOREM 7. *The notation being as above, W_u is the intersection of V and the hyperplane H_u defined by the equation*

$$u_1 X_1 + \ldots + u_n X_n - u_{n+1} = 0.$$

If $\dim V \geqq 2$, *then* $V \cap H_u = W_u$ *is a variety, i.e. the extension $k_u(x)$ is a regular extension of k_u.*

Proof: From the inclusions

$$k(u_1, \ldots, u_n) \subset k(u_1, \ldots, u_{n+1}) \subset k_u(x)$$

and the fact that u_{n+1} is transcendental over $k(u_1, \ldots, u_n)$ it follows that $k_u(x)$ has dimension $r - 1$ over k_u. Hence W_u has dimension $r - 1$. Since it has a generic point contained in $V \cap H_u$, it is itself contained in $V \cap H_u$. Let (x') be any point in $V \cap H_u$. Since it is in V, there is a specialization $(x) \to (x')$ over $k(u_1, \ldots u_n)$. Since it is in H_u, we can write $u_{n+1} = u_1 x_1' + \ldots + u_n x_n'$ and hence u_{n+1} remains fixed under our specialization. This proves that our specialization is actually over k_u, and hence that (x') lies in W_u.

There remains to prove that if $\dim V \geqq 2$, then the extension $k_u(x)$ over k_u is regular. This will be a consequence of the following lemma, due to Zariski in characteristic 0 and extended by Matsusaka to characteristic p.

Lemma. *Let K be a regular finitely generated extension of an infinite field k, and let y, z be two elements of K algebraically independent over k such that z is not in $K^p k$. Then for all but a finite number of constants c of k, K is a regular extension of $k(y + cz)$.*

Proof: We shall use throughout the fact that a subfield of a finitely generated extension is also finitely generated (Proposition 6 of Chapter III, § 2).

If w is an element of K, and if there exists a derivation D of K over k such that $Dw \neq 0$, then K is separable over $k(w)$, by the Corollary to Proposition 2, Chapter VII, § 1. For all elements $c \in k$, except possibly one, we have $D(y + cz) = Dy + cDz \neq 0$. In what follows, we assume that the constants $c_1, c_2, \ldots$, are different from this exceptional constant, and hence that K is separable over $k(y + c_1z)$ or $k(y + c_2z)$.

Denote $k(y + c_iz)$ by E_i, and let E_i' be the algebraic closure of E_i in K. We must show that $E_i' = E_i$ for all but a finite number of constants. Note that $k(y, z) = E_1E_2$ is the compositum of E_1 and E_2, and that it has transcendence degree 2 over k. Hence E_1' and E_2' are free over k. Being subfields of a regular extension of k, they are regular over k, and are therefore linearly disjoint by Theorem 3 of Chapter III, § 1.

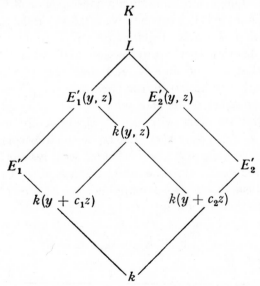

By construction, E_1' and E_2' are finite separable algebraic extensions of E_1 and E_2 respectively. Let L be the separable algebraic closure of $k(y, z)$ in K. There is only a finite number of intermediate fields between $k(y, z)$ and L. Furthermore, by Proposition 1 of Chapter III, § 1 the fields $E_1'(y, z)$ and $E_2'(y, z)$ are linearly disjoint over $k(y, z)$. Let c_1 range over the finite number of constants which will exhaust the intermediate extensions between L and $k(y, z)$ obtainable by lifting over $k(y, z)$ a field of type E_i'. If c_2 is now chosen different from any one of these constants c_1, the only way in which the condition of linear disjointness mentioned above can be compatible with our choice of c_2 is that $E_2'(y, z) = k(y, z)$, i.e. that $E_2' = k(y + c_2 z)$. This means that $k(y + c_2 z)$ is algebraically closed in K, and hence that K is regular over $k(y + c_2 z)$. This proves our lemma.

The notations being the same as in the lemma, suppose that t is transcendental over K. Put $K_t = K(t)$ and $k_t = k(t)$. Then K_t is a regular extension of $k_t(y + tz)$. Proof: Let $t_1, t_2, \ldots$, be infinitely many quantities algebraically independent over k, and let $k^* = k(t_1, t_2, \ldots)$, $K^* = K(t_1, t_2, \ldots)$ be the fields obtained by adjoining them to k and K respectively. By the lemma, K^* is regular over $k^*(y + t_i z)$ for all but a finite number of integers i, say for $i = 1$. Our assertion is then a consequence of Corollary 6, Theorem 3, Chapter III, § 1, where we lift $K(t_1)$ and $k(t_1)$ over the field k^* which is linearly disjoint from $K(t_1)$ over $k(t_1)$.

In the application to the intersection of V with H_u, we may assume say that x_n is not in $K^p k$, and then take $y = u_1 x_1 + \ldots + u_{n-1} x_{n-1}$ and $z = x_n$, so that $u_{n+1} = y + u_n z$. This concludes the proof that $k_u(x)$ is regular over k_u.

Suppose that we start with $n + 1$ algebraically independent quantities $t_1, \ldots, t_{n+1}$ over $k(x)$. Let H_t be the hyperplane defined by $t_1 X_1 + \ldots + t_n X_n - t_{n+1} = 0$. It is called a *generic hyperplane over k*. We are interested in the intersection of V with H_t. There is an isomorphism $\sigma: k(u) \to k(t)$ mapping u_i on $t_i (i = 1, \ldots, n + 1)$ such that σ is identity on k. We have

$$(V \cap H_u)^\sigma = V^\sigma \cap H_u^\sigma = V \cap H_t$$

and we see therefore that $V \cap H_t$ is a $k(t)$-variety, and is a variety if dim $V \geqq 2$. If we extend σ to an isomorphism of the field $k(u, x)$, and let (x^*) be the image of (x) under σ, then (x^*) is a generic point of $V \cap H_t$ over $k(t)$. Thus it is clear conversely that any generic point (y) of $V \cap H_t$ over $k(t)$ is also a generic point of V over $k(t_1, \ldots, t_n)$, and hence over k. In addition to that, we have also the following useful statement.

PROPOSITION 10. *Let V be defined over k. Let H_u be a generic hyperplane over k, and let $W_u = V \cap H_u$. Let (x) and (y) be two distinct independent generic points of W_u over $k(u)$. Then they are independent generic points of V over k.*

Proof: This will be proved by counting dimensions in the following diagram:

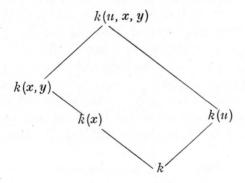

As we have seen, (x) is a generic point of V over k, and hence $\dim_k(x) = r$. We also have $\dim_k(u) = n + 1$. Since (x) and (y) are independent points of W_u over $k(u)$, we note that the dimension of $k(u, x, y)$ over $k(u)$ is equal to $2(r - 1)$. Hence $\dim_k(u, x, y) = n + 1 + 2(r - 1) = n + 2r - 1$.

Since we have

$$u_{n+1} = u_1 x_1 + \ldots + u_n x_n$$
$$u_{n+1} = u_1 y_1 + \ldots + u_n y_n$$

we subtract these equations, and get $u_1(x_1 - y_1) + \cdots$

$+ u_n(x_n - y_n) = 0$. By hypothesis, $(x) \neq (y)$, and from this we see that $k(u, x, y)$ can be obtained from $k(x, y)$ by adjoining $u_2, \ldots, u_n$ (if say $x_1 \neq y_1$).   Hence the dimension of $k(u, x, y)$ over $k(x, y)$ is at most $n - 1$. This implies that the dimension of $k(x, y)$ over $k(x)$ must be at least r, and hence that (x), (y) are independent generic points of V over k. This proves our proposition.

Remark that we have not assumed that dim $V \geqq 2$, and that (x), (y) may well have dimension 0 over $k(u)$, i.e. they may be algebraic over $k(u)$, in which case they are conjugate over $k(u)$, and we can then give the following additional information.

PROPOSITION 11.   *Suppose that V has dimension* 1.   *Let* H_u *be a generic hyperplane over k, and let* (x) *be a point in* $V \cap H_u = W_u$. *Then* $k_u(x)$ *is separable algebraic over* k_u.

Proof: By Proposition 4 of Chapter III, § 1 we may assume that one of the x_i, say x_1, is a separating transcendence base of $k(x)$ over k.   In that case, we can find polynomials $f_i(X_1, X_i) (i = 2, \ldots, n)$ such that $f_i(x_1, x_i) = 0$ but $\partial f_i / \partial x_i \neq 0$.   The matrix $\|\partial f_i / \partial X_j\|$ $(i, j = 2, \ldots, n)$ has therefore rank $n - 1$. Adding the equation $u_1 X_1 + \ldots + u_n X_n - u_{n+1}$ which vanishes on (x), we see that we can find n polynomials with coefficients in $k(u)$ vanishing on (x) and satisfying the condition of Theorem 3, Chapter VII, § 1.   This proves that (x) is separable algebraic over $k(u)$.

Before going further, let us strengthen part of Theorem 7.

PROPOSITION 12. *Let* $\mathfrak{p}$ *be the prime ideal of V in* $k[X]$. *Then the ideal* $(\mathfrak{p}, H_u)$ *generated by* $\mathfrak{p}$ *and the generic hyperplane* H_u *in* $k_u[X]$ *is the prime ideal belonging to* $V \cap H_u$.

Proof: Let $f \in k_u[X]$ vanish on $V \cap H_u$. After multiplying f by a suitable polynomial in $k[u]$ we may assume that all coefficients of f are in $k[u]$. Since H_u contains u_{n+1} to the first power and with coefficient 1, we can find a polynomial $g(u, X) \in k[u, X]$ such that $f - H_u g(u, X)$ does not contain u_{n+1}, or in other words lies in $k[u_1, \ldots, u_n, X]$. Since $f(x) = 0$ and $H_u(x) = 0$ and since (x) is a generic point of V over $k(u_1, \ldots, u_n)$, it follows that $f - H_u g(u, X)$ lies in the ideal generated by $\mathfrak{p}$ in $k(u_1, \ldots, u_n)[X]$ by Corollary 1

· of Theorem 8, Chapter III, § 2. This proves our proposition.

We conclude this chapter by showing how the singular points of V and its generic hyperplane section determine each other.

PROPOSITION 13. *Let V be of dimension ≥ 2 and let H_u be a generic hyperplane over k, as before. Then a point of $W_u = V \cap H_u$ is simple on W_u if and only if it is simple on V.*

Proof: Let (y) be a point of W_u which is simple on V. By definition, we can find $n - r$ polynomials $f_i(X) \epsilon k[X]$ $(i = 1, \ldots, n - r)$ vanishing on V, and such that the matrix

$$\begin{pmatrix} \dfrac{\partial f_1}{\partial y_1} & \cdots & \dfrac{\partial f_1}{\partial y_n} \\ \cdots\cdots\cdots\cdots \\ \dfrac{\partial f_{n-r}}{\partial y_1} & \cdots & \dfrac{\partial f_{n-r}}{\partial y_n} \end{pmatrix}$$

has rank $n - r$. After a suitable renumbering of the variables, we may assume that the submatrix of the partials taken with respect to $y_1, \ldots, y_{n-r}$ has that rank. To show that (y) is simple on W_u, it will suffice to prove that the Jacobian matrix taken with the f_i and H_u has rank $n - (r - 1) = n - r + 1$ when evaluated at (y). Suppose that it does not. Then for $j = n - r + 1, \ldots, n$ all the following determinants vanish:

$$\begin{vmatrix} \dfrac{\partial f_1}{\partial y_1} & \cdots & \dfrac{\partial f_1}{\partial y_{n-r}} & \dfrac{\partial f_1}{\partial y_j} \\ \cdots\cdots\cdots\cdots\cdots\cdots \\ \dfrac{\partial f_{n-r}}{\partial y_1} & \cdots & \dfrac{\partial f_{n-r}}{\partial y_{n-r}} & \dfrac{\partial f_{n-r}}{\partial y_j} \\ u_1 & \cdots & u_{n-r} & u_j \end{vmatrix}$$

This implies that $u_j (j = n - r + 1, \ldots, n)$ lies in the field $k(y, u_1, \ldots, u_{n-r})$. Since (y) lies in V, its dimension over k is at most r, and we have shown that the dimension of (u) over $k(y)$ is at most $n - r$. On the other hand, consider the following diagram:

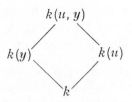

We have $\dim_k(u) = n + 1$. Let $\dim_k(y) = s$, and $\dim_{k(u)}(y) = s'$. Then the dimension of (u) over $k(y)$ is equal to $n + 1 + s' - s$. Since (y) is in V, we have $s \leqq r$, and hence

$$n + 1 + s' - s \geqq n + 1 + s' - r \geqq n + 1 - r$$

because $s' \geqq 0$. This is a contradiction, and we have proved that (y) is simple on W_u.

Conversely, suppose that (y) is singular on V, and is contained in W_u. Since the prime ideal belonging to W_u is equal to $(\mathfrak{p}, H_u)$ by the preceding proposition, it follows that the rank of the Jacobian matrix of (y) relative to W_u can be at most one greater than the rank of the Jacobian matrix of (y) relative to V. Hence (y) cannot be simple on W_u. This concludes the proof of the proposition.

We can give another interpretation to the above arguments. One usually defines the *tangent linear variety* at a simple point (y) of V to be the variety consisting of the zeros of the linear equations

$$\frac{\partial f}{\partial y_1} (X_1 - y_1) + \ldots + \frac{\partial f}{\partial y_n} (X_n - y_n) = 0$$

as f ranges over all polynomials in the prime ideal belonging to V (or equivalently, over a basis of this prime ideal). We say that a hyperplane is *tangent* to V at (y) if it contains the tangent linear variety. The next proposition is then an immediate consequence of the preceding arguments.

PROPOSITION 14. *Let V be defined over k. Let H_u be a generic hyperplane over k. Let (y) be any point of V. Then H_u is not tangent to V at (y).*

Literature

The theory of local rings originated with Krull [2] and the application of that theory to simple points is due to Zariski [4], who discovered that the local ring of a simple point is regular, this property being characteristic if the ground field is perfect. The only important theorem concerning the local ring of a simple point which we have not proved here is that it is a unique factorization domain. The reader will find a proof in Zariski [4]. The proof uses the techniques of complete local rings, and hence lies outside the global techniques to which we have restricted ourselves in this book. No other proof seems to be known at the time this book is being written.

The problem concerning the existence of a complete, or projective non-singular model of a function field is still open. The local uniformization proved by Zariski in characteristic 0 [3] has just been extended to characteristic p by Abhyankar in the case of surfaces [1].

ABHYANKAR, S.
[1] Local uniformization on algebraic surfaces over ground fields of characteristic $p \neq 0$, Annals of Math., Vol. **63**, No. 3, May 1956, pp. 491—526.

KRULL, W.
[2] Dimensionstheorie in Stellenringe, J. Reine u. Angew. Math., Vol. **179**, **1938**, pp. 204—226.

ZARISKI, O.
[3] Local uniformization on algebraic varieties, Annals of Math., Vol. **41**, **1940**, pp. 852—896.
[4] The concept of a simple point of an abstract algebraic variety, Trans. Am. Math. Soc., Vol. 62, No. 1, July 1947, pp. 1—52.

CHAPTER IX

Algebraic Groups

We do not give here any comprehensive theory, but only a few striking results, especially concerning abelian varieties and taken from Weil's book on the subject. Thanks to recent papers, there now exists in the literature a complete self-contained and elegant exposition of the foundations of the theory of algebraic groups.

For simplicity we limit ourselves to connected groups.

A *group variety* (or a connected algebraic group) is an abstract variety G together with a group structure such that the law of composition

$$f : G \times G \to G$$

with $f(x, y) = xy$ is an everywhere defined rational map of $G \times G$ into G, and the inverse mapping $\varphi(x) = x^{-1}$ is an everywhere defined rational map of G into G. The group G is said to be defined over k if the abstract variety, the law of composition, and the inverse are defined over k. The unit element e of G is then rational over k, for we can write $e = xx^{-1}$, whence e is rational over $k(x)$ for a generic point x of G over k. Taking two independent generic points x and y, we see that e is rational over $k(x) \cap k(y) = k$.

THEOREM 1. *If G is a group variety and a is an element of G then the mapping $x \to ax$ is a biholomorphic transformation of G onto itself, which is called the left translation by a.*

Proof: If we put $T_a(x) = f(a, x) = ax$, and if we put $T'_a(x) = a^{-1}x$ then one sees immediately that T_a and T'_a are everywhere-defined rational maps of G into itself and that $T_a T'_a$ is the identity. Hence T_a is biholomorphic.

COROLLARY. *A group variety is non-singular*

Proof: If a is an arbitrary point, and b is a simple point, then $T_{ab^{-1}}$ maps b onto a. Hence a is simple.

We shall make a few remarks on the composition law. Let x

be a generic point of G over k. Let $\mathfrak{o}$ be the local ring of the origin e in $k(x)$. Let $t_1, \ldots, t_r$ be a set of generators for its maximal ideal $\mathfrak{m}$ (r being the dimension of G). Then each t_i can be written $t_i = f_i(x)$ where f_i is a function on G. Let y be another generic point of G over k, independent of x, and put $u_i = f_i(y)$. Then (t, u) form a system of parameters for the point (e, e) on $G \times G$. Since the functions $f_i(xy)$ are obviously elements of the local ring $\mathfrak{O}$ of (e, e) in $k(x, y)$ and since (e, e) is rational over k, we can write

$$f_i(xy) \equiv a_1 t_1 + \ldots + a_r t_r + b_1 u_1 + \ldots + b_r u_r (\mathrm{mod}\,\mathfrak{M}^2)$$

where a_i, b_j lie in k and where $\mathfrak{M}$ is the maximal ideal of $\mathfrak{O}$. On the other hand, if we specialize y to e, then $f_i(e) = 0$, and this shows that $a_j = 0$ for $j \neq i$. Similarly, $b_j = 0$ for $j \neq i$. Hence we get

$$f_i(xy) \equiv f_i(x) + f_i(y) \ (\mathrm{mod}\,\mathfrak{M}^2).$$

This means that to the first approximation, the composition law on a group variety is linear and commutative in a neighborhood of the origin. It also shows that $f_i(x^n) \equiv nf_i(x) \bmod \mathfrak{m}^2$, whence

PROPOSITION 1. *If x is a generic point of G over k, and n is not divisible by the characteristic, then $x^n \neq 0$, and the rational map $x \to x^n$ is generically surjective.*

Proof: The first statement is clear. As to the second, the r quantities $f_i(x^n)\,(i = 1, \ldots, r)$ obviously form a system of parameters of $\mathfrak{m}$ mod $\mathfrak{m}^2$, and hence are algebraically independent over k, say by the characterization of parameters in terms of a Jacobian. They are contained in $k(x^n)$, which must therefore have dimension r over k, and thus x^n is a generic point of G over k.

Examples of algebraic groups abound. For instance:

The group of all non-singular $n \times n$ matrices, which is called the linear group, the composition law being ordinary multiplication of matrices, and the identity being the unit matrix.

The group of projective transformations of projective n-space (called the projective group).

The group of all $n \times n$ matrices with determinant 1, and the group of all triangular non-singular matrices.

The affine line S^1 under addition.

The affine line from which the origin is removed, with multiplication as the law of composition.

It will be noted that the varieties of these groups are not complete. We shall study below complete group varieties, which are of an entirely different type. For an example, see the section dealing with the construction of the Jacobian variety of a curve.

A homomorphism $f : G \to H$ of a group variety G into a group variety H is said to be a *rational homomorphism* if f is at the same time an everywhere defined rational map. It is called a *rational isomorphism* if in addition it is an isomorphism for the group structure. It need not be biholomorphic, but must be purely inseparable (as a rational map). For instance, if x is a generic point of G over k, and $x^{(p)}$ denotes the point obtained from x by applying the automorphism of the universal domain raising every element to the pth power, we get such a rational isomorphism of G onto a group variety $G^{(p)}$. If a rational isomorphism is also a birational map, then it is biholomorphic (say by ZMT), and is called a *birational isomorphism*.

REMARK 1. *Let $f : G \to H$ be a rational map of G into H such that for two independent generic points x and y of G over k we have $f(xy) = f(x)f(y)$. Then f is everywhere defined, and for any two points a, b of G we have $f(ab) = f(a)f(b)$.*

Proof: Let a be a point of G. We may assume a rational over k. Let now x, y be independent generic points of G over k. We can write $f(x) = f(xy)f(y)^{-1}$. Then ay is a generic point of G over k, and if we specialize (x, y) to (a, y), the expression $f(ay)f(y)^{-1}$ is defined. Hence $f(a)$ is defined. The rest is clear.

REMARK 2: *The notation being as in Remark 1, the mapping f sends G onto a closed subset of H, which is itself a group variety.*

Proof: The set-theoretic image of G is a subgroup H' of H which contains a non-empty open set on H by Proposition 4 of Chapter IV, § 2. Hence the cosets of H' are contained in a proper closed subset of H, which must be empty by Theorem 1.

The above two remarks give examples of a situation in which

we are able to deduce a global behavior of groups and homomorphisms from the behavior given only at generic points. We shall now describe how the structure of a group itself can be fully recovered from its structure at generic points.

Let V be a variety defined over a field k. By a *normal law of composition* on V, one means a generically surjective rational map $f : V \times V \to V$, which satisfies the following conditions:

G1. If u, v are two independent generic points of V over k, and $w = f(u, v)$ then $k(u, v) = k(u, w) = k(v, w)$ so that in particular any two points among u, v, or w are independent generic points of V over k.

G2. f is associative at generic points, i.e.

$$f(f(u, v), w) = f(u, f(v, w)).$$

To simplify the notation, we write $u.v$ instead of $f(u, v)$.

If V' is any variety birationally equivalent to V over k, say $T : V \to V'$ is a birational correspondence, then we can obviously define a normal law of composition on V' by transferring that of V, namely by putting $T(u) . T(v) = T(w)$. One can then prove the following result: *There exists a group variety G defined over k and birationally equivalent to V over k such that the group law on G is that obtained by transferring that of V. The uniqueness of the group G then follows from Remark 1.* For the proof, we refer the reader to Weil [16], [17], [18].

One should also note that transformation spaces and homogeneous spaces can be defined in a manner analogous to that of groups, by using the classical topological definition, with the additional requirement that the spaces be varieties, and that the maps be everywhere defined rational maps.

All the examples of algebraic groups given above are *linear*, i.e. there exists a rational isomorphism of those groups into the linear group. The underlying varieties are all biholomorphic to affine varieties, because the linear group itself has this property. (If t_{ij} are the variables of the linear group, $(i, j = 1, \ldots, n)$, and if $D(t)$ is the determinant det (t_{ij}) then an affine generic point is given by $(t, 1/D(t))$.)

It is more difficult to construct examples of group varieties whose underlying variety is complete. The rest of this chapter will be devoted to proving the basic facts concerning such group varieties, which are called *abelian varieties*. The terminology is in part justified by the following theorem of Chevalley.

THEOREM 2. *The law of composition on a complete group variety G is commutative.*

Proof: If k is an algebraically closed field of definition for G, and x, y two independent generic points of G over k, it suffices to prove that $xy = yx$. Let W be the locus of (x, yxy^{-1}) over k. Specializing y to e shows that W contains the diagonal on $G \times G$. If we can show that $\dim W \leq r$ (the dimension of G) then W will be equal to the diagonal and we are through. On the product $G \times G$ we intersect W with the variety $e \times G$ which is of dimension r. If (e, a) is a point of this intersection, then using the fact that G is complete, there exists a point b of G such that the specialization $(x, yxy^{-1}) \to (e, a)$ can be extended to a specialization of $y \to b$. We must then have $a = beb^{-1} = e$. This proves that the only point in the intersection is the origin (e, e) which has dimension 0. Using the dimension theorem for simple points, we must have

$$0 \geq \dim W + r - 2r$$

whence $\dim W \leq r$. This concludes the proof.

The first fundamental theorem concerning algebraic groups is the following:

THEOREM 3. *Let $f : V \to G$ be a rational map of a variety V (say affine) into a group variety G. If a is a simple point of V at which f is not defined, then there exists a subvariety W of V of codimension 1 containing a such that f is not defined at W.*

Proof: Let k be a field of definition for f, V, and G, and such that a is rational over k. Let x, y be two independent generic points of V over k. We define a rational map $F : V \times V \to G$ by putting $F(x, y) = f(x)f(y)^{-1}$. We contend that f is defined at a if and only if F is defined at (a, a). This is trivial in one direction. In the other, suppose F is defined at (a, a). Then F is *a fortiori* defined at (a, y) because (a, a) is a specialization of (a, y). We can write $f(x) = f(y)F(x, y)$. The expression on the right-hand side is defined at (a, y), and hence f is defined at a.

If F is defined at (a, a), then it takes the value e in G. Let G_0 be an affine representative of G on which e has a representative e_0. Let z be the affine representative on G_0 of the point $F(x, y)$. If F is not defined at (a, a), then at least one coordinate of z is not defined at (a, a). Let s be the dimension of V. By Proposition 3 of Chapter VI, § 1 this implies that this coordinate is not defined at a subvariety U of $V \times V$ having dimension $2s - 1$ and containing (a, a). The intersection of U with the diagonal on $V \times V$ contains (a, a) which is a simple point on $V \times V$. By the dimension theorem for simple points, a component W^* of this intersection containing (a, a) has dimension $\geq 2s - 1 + s - 2s = s - 1$. Let (w, w) be a generic point of W^* over a suitable field. Then F cannot be defined at (w, w) (i.e. at W^*) and hence by the remarks at the beginning of the proof, f cannot be defined at the locus W of w over k. This W is clearly the desired subvariety of V.

COROLLARY. *If $f : V \to A$ is a rational map of a variety V into an abelian variety A, then f is defined at every simple point of V.*

Proof: Use the corollary to Theorem 1 of Chapter VI, § 1 and the preceding theorem.

LEMMA. *Let A be an abelian variety, V and W two arbitrary varieties, and P a simple point of V. Let $f(x, y)$ be a rational map of $V \times W$ into A. Assume that f is defined on $P \times W$, and is constant on this subvariety of $V \times W$. Then there is a rational map f_0 of V into A such that $f(x, y) = f_0(x)$.*

Proof: We shall prove the lemma by induction on the dimension of W. Suppose first W has dimension 1. Since the statement of our theorem is birational in W, we may assume W is a complete non-singular curve. Let k be an algebraically closed field of definition for all the above objects, and such that P is rational over k. Let (x) and (y) be two independent generic points of V and W over k. Let $z = f(x, y)$. We must show that z is rational over $k(x)$. Let T be the locus of (x, z) over k. T is a correspondence between V and A. Since A is complete, the specialization $(x) \to P$ can be extended to $(x, z) \to (P, z')$, and by the dimension theorem for correspondences (or for simple points) a component of $T \cap (P \times A)$ has dimension at least equal to that of z over $k(x)$. If z is not rational over $k(x)$, this dimension is at least 1. We may suppose (P, z') is the generic point of such a component. Our specialization can be extended to a specialization (y') of (y). Since (P, y') is simple on $V \times W$, and f is defined at every simple point, f is constant at (P, y'). This is a contradiction.

Now we reduce our lemma to the case where W has dimension 1. If as before (x) and (y) are two independent generic points of V and W over k, we can interpose a sequence of fields $k \subset K_1 \subset K_2 \subset \ldots \subset K_s = k(y)$ between k and $k(y)$ such that K_i has dimension 1 over K_{i-1}, and K_{i-1} is algebraically closed in K_i. After making suitable purely inseparable extensions of $k(y)$, we may assume that each extension K_i of K_{i-1} is regular. Going down stepwise from the top, we consider K_s over K_{s-1} as the function field of a curve W_s and we have a rational map of $V \times W_s$ into A also defined by $z = f(x, y)$. It is trivial from our hypotheses that f is constant on $P \times W_s$. Hence z is rational over $k(x) K_{s-1} \cdot K_{s-1}$ is the function field over k of a variety W', of dimension one less than that of W. We can apply our induction hypothesis to conclude that z is rational over $k(x)$. This proves our lemma.

THEOREM 4. *Let A be an abelian variety, and V, W two arbitrary varieties. Let $f : V \times W \to A$ be a rational map, defined over a field k. Then there exist rational maps f_1, f_2 of V and W respectively into A such that $f(x, y) = f_1(x) + f_2(y)$. Furthermore, f_1 and f_2 are uniquely determined up to a constant by this relation.*

Proof: Let P be a simple point of V. Let k be a field of definition for all the above objects, over which P is rational. Since f is defined on $P \times W$ (which is simple on $V \times W$), we can write $f(P, y) = f_1(y)$, where f_1 is a rational map of W into A. Using the lemma, we can write

$$f(x, y) - f(P, y) = f_2(x),$$

where f_2 is a rational map of V into A. This gives us the decomposition of f. The uniqueness of f_1 and f_2 is obvious.

THEOREM 5. *Let $f : G \to A$ be a rational map of a group variety into an abelian variety. Then f is a homomorphism, up to a translation.*

Proof: We denote by e the unit element of G, and by 0 that of A. After a suitable translation (by $f(e)$) we may assume that $f(e) = 0$. We shall then prove that f is a homomorphism. Let F be the rational map of $G \times G$ into A defined by $F(x, y) = f(x\,y)$. By the preceding theorem, we can write

$$f(xy) = F(x, y) = f_1(x) + f_2(y),$$

and after adjusting f_2 by a constant, we may assume that $f_2(e) = 0$. Putting $y = e$ in the above equation, we get $f_1(x) = f(x)$. It is then clear that $f_2(y)$ is $f(y)$, and this proves our theorem.

COROLLARY. *If f is a rational map of a variety birationally equivalent to affine space into an abelian variety then f is constant.*

Proof: It suffices to assume our variety is equal to affine space S^n. This is the product of n affine lines. Using Theorem 4, it suffices to prove our theorem for S^1. By Theorem 5, f must be a homomorphism both for the additive group law on S^1, and for the multiplicative group law on S^1 from which we delete the origin. The reader will immediately verify that this can occur only if f is constant.

REMARK: One may say that a variety V is *pure* if there is a field of definition k for V such that the function field $k(V)$ of functions on V defined over k is a purely transcendental extension of k. In that case, one says that V is *k-pure*, or *pure over k*. If the function field $k(V)$ is contained in a purely transcendental extension of k, then we say V is *k-semi-pure*, or *semi-pure over k*. One says it is *semi-pure* if there exists such a field of definition. From the above corollary, one deduces immediately that a rational map of a semi-pure variety into an abelian variety is constant. Note that the rational map may be defined over a field k such that V is neither pure nor semi-pure over k, but becomes semi-pure over the algebraic closure of k, for instance. In other words, the proposition is "geometric", it is independent of the field of definition.

Literature

The theory of algebraic groups has now acquired a very great importance in algebraic geometry. The foundations for this theory are given in Weil [16],[17],[18] and Rosenlicht [14]. These papers give a self-contained exposition. One should also mention Barsotti's paper [1]. The fundamental theorem concerning the structure of group varieties states that G contains a unique maximal linear subgroup L such that the factor group variety G/L is an abelian variety. For a proof see [1] and [14].

The theory of linear groups in abstract algebraic geometry was originated by Chevalley [3] and Kolchin [8], [9], and continued by Borel [2]. The deepest results in this direction are due to Chevalley, who first exhibits the analogues to all the classical Lie groups in the abstract case [4] and then shows that these give all simple linear algebraic groups [5].

In another direction, the theory of abelian varieties was originated (in the abstract case) by Weil [15] and has been greatly enriched by results of Chow [6] and Matsusaka [12]. Rosenlicht's generalized Jacobians [13] form a bridge between abelian varieties and linear groups. Their analogues in higher dimension are not yet known at the time this book is being written. Igusa [7] has given interesting results on these Jacobians.

Finally, arithmetic applications of the theory of algebraic groups have been given in [10], [11] where it is shown how class field theory is intimately related with them.

BARSOTTI, I.
[1] Structure theorems for group varieties, Annali di Matematica pura ed applicata, Serie IV, T. 38 (1955), pp. 77—120.

BOREL, A.
[2] Groupes linéaires algébriques, Annals of Math., Vol. 64, July 1956, pp. 20—82.

CHEVALLEY, C.
[3] Théorie des groupes de Lie (Groupes algébriques), Hermann et Cie., Paris, 1951.
[4] Sur certains groupes simples, Tohoku Math. Jornal, Vol. 7, No. 1—2, pp. 14—66.
[5] Seminar Ecole Normale Supérieure, Paris, 1957—1958.

CHOW, W. L.
[6] Forthcoming papers on the Picard and Albanese varieties, to appear in the Annals of Math.

Igusa, J.

[7] Fibre systems of Jacobian varieties, Am. J. of Math., Vol. 78, No. 1, January 1956, pp. 171—199.

Kolchin, E.

[8] On certain concepts in the theory of algebraic matric groups, Annals of Math., Vol. 49, No. 4, October 1948, pp. 774—789.

[9] Algebraic matric groups and the Picard Vessiot theory..., Annals of Math., Vol. 49, No. 1, January 1948, pp. 1—42.

Lang, S.

[10] Algebraic groups over finite fields, Am. J. of Math.,Vol. 78, No. 3, July 1956, pp. 555—563.

[11] Unramified class field theory over function fields in several variables, Annals of Math., Vol. 64, No. 2, September 1956, pp. 285—325.

Matsusaka, T.

[12] Algebraic construction of the Picard variety I and II, Jap. Journal of Math., Vol. 21 and 22, pp. 217—235 and 51—62.

Rosenlicht, M.

[13] Generalized Jacobian varieties, Annals of Math., Vol. 59, May 1954, pp. 505—530.

[14] Some basic theorems on algebraic groups, Am. J. of Math., Vol. 78, No. 2, April 1956, pp. 401—443.

Weil, A.

[15] Variétés abéliennes et courbes algébriques, Hermann et Cie., Paris, 1948.

[16] Algebraic groups of transformations, Am. J. of Math., Vol. 77, April 1955, pp. 355—391.

[17] Algebraic groups and homogeneous spaces, Am. J. of Math., Vol. 77, No. 3, July 1955, pp. 493—512.

[18] On the field of definition of a variety, Am. J. of Math., Vol. 78, No. 3, July 1956, pp. 509—524.

CHAPTER X

Riemann-Roch Theorem

We shall give Weil's proof of the Riemann-Roch theorem. The construction of the Jacobian variety of a curve is a consequence of the theorem, and of general theorems on algebraic groups, and can be read immediately after Theorem 7. We have sketched this construction, partly in order to give the reader an idea of the nature of complete group varieties. It can be shown that any complete group variety is in fact a homomorphic image of a Jacobian variety.

We end the chapter, and the book, with an amusing application of the Riemann-Roch theorem, namely Harnack's theorem.

1. Lemmas on valuations

We shall refine Proposition 7 of Chapter I, § 6. For this we need what is commonly known as the *approximation theorem*, which gives us the existence of a function having given zeros and poles. We shall deal with an arbitrary field K, and a finite set of discrete valuations belonging to valuation rings $\mathfrak{o}_1, \ldots, \mathfrak{o}_n$ of K. As we know, we can then speak of the order of an element of K at $\mathfrak{o}_i$, this order being an integer, and we can also speak of zeros and poles. The next results will of course be applied to the concrete case of a finite number of divisors on a curve, but it is just as simple to give the proof in a slightly more abstract situation.

PROPOSITION 1. *If $\mathfrak{o}_1$ and $\mathfrak{o}_2$ are two discrete valuation rings in a field K, such that $\mathfrak{o}_1 \subset \mathfrak{o}_2$, then $\mathfrak{o}_1 = \mathfrak{o}_2$.*

Proof: We shall first prove that if $\mathfrak{p}_1$ and $\mathfrak{p}_2$ are their maximal ideals then $\mathfrak{p}_2 \subset \mathfrak{p}_1$. Let $y \in \mathfrak{p}_2$. If $y \notin \mathfrak{p}_1$, then $1/y \in \mathfrak{o}_1$, whence $1/y \in \mathfrak{o}_2$, a contradiction. Hence $\mathfrak{p}_2 \subset \mathfrak{p}_1$. Every unit of $\mathfrak{o}_1$ is *a fortiori* a unit of $\mathfrak{o}_2$. An element y of $\mathfrak{p}_2$ can be written $y = \pi_1^{\nu_1} u$ where u is a unit of $\mathfrak{o}_1$ and π_1 is an element of order 1 in $\mathfrak{p}_1$. If π_1

is not in $\mathfrak{p}_2$ it is a unit in $\mathfrak{o}_2$, a contradiction. Hence π_1 is in $\mathfrak{p}_2$, and hence so is $\mathfrak{p}_1 = \mathfrak{o}_1 \pi_1$. This proves $\mathfrak{p}_2 = \mathfrak{p}_1$. Finally, if u is a unit in $\mathfrak{o}_2$, and is not in $\mathfrak{o}_1$, then $1/u$ is $\mathfrak{p}_1$, and thus cannot be a unit in $\mathfrak{o}_2$. This proves our proposition.

From now on, we assume that *our valuation rings* $\mathfrak{o}_i$ $(i = 1, \ldots, n)$ *are distinct, and hence have no inclusion relations.*

PROPOSITION 2. *There exists an element* y *of* K *having a zero at* $\mathfrak{o}_1$ *and a pole at* $\mathfrak{o}_j (j = 2, \ldots, n)$.

Proof: This will be proved by induction. Suppose $n = 2$. Since there is no inclusion relation between $\mathfrak{o}_1$ and $\mathfrak{o}_2$, we can find $y \in \mathfrak{o}_2$ and $y \notin \mathfrak{o}_1$. Similarly, we can find $z \in \mathfrak{o}_1$ and $z \notin \mathfrak{o}_2$. Then y/z satisfies our requirements.

Now suppose we have found an element y of K having a zero at $\mathfrak{o}_1$ and a pole at $\mathfrak{o}_2, \ldots, \mathfrak{o}_{n-1}$. Let z be such that z has a zero at $\mathfrak{o}_1$ and a pole at $\mathfrak{o}_n$. Then for sufficiently large r, $y + z^r$ satisfies our requirements, because we have schematically zero plus zero $=$ zero, zero plus pole $=$ pole, and the sum of two elements of K having poles of different order again has a pole.

A high power of the element y of Proposition 2 has a high zero at $\mathfrak{o}_1$ and a high pole at $\mathfrak{o}_j (j = 2, \ldots, n)$. Adding 1 to this high power, and considering $1/(1 + y^r)$ we get

COROLLARY. *There exists an element* z *of* K *such that* $z - 1$ *has a high zero at* $\mathfrak{o}_1$, *and such that* z *has a high zero at* $\mathfrak{o}_j (j = 2, \ldots, n)$.

Denote by ord_i the order of an element of K under the discrete valuation associated with $\mathfrak{o}_i$. We then have the following approximation theorem.

THEOREM 1. *Given elements* $a_1, \ldots, a_n$ *of* K, *and an integer* N, *there exists an element* $y \in K$ *such that* $\mathrm{ord}_i (y - a_i) > N$.

Proof: For each i, use the corollary to get z_i close to 1 at $\mathfrak{o}_i$ and close to 0 at $\mathfrak{o}_j$ $(j \neq i)$, or rather at the valuations associated with these valuation rings. Then $z_1 a_1 + \ldots + z_n a_n$ has the required property.

In particular, we can find an element y having given orders at the valuations arising from the $\mathfrak{o}_i$. This is used to prove the following inequality.

COROLLARY. *Let E be a finite algebraic extension of K. Let Γ be the value group of a discrete valuation of K, and Γ_i the value groups of the finite number of inequivalent discrete valuations of E extending that of K. Let e_i be the ramification index of Γ in Γ_i. Then $\sum e_i \leq [E : K]$.*

Proof: Select elements

$$y_{11}, \ldots, y_{1e_1}, \ldots, y_{r1}, \ldots, y_{re_r}$$

of E such that $y_{iv}(v = 1, \ldots, e_i)$ represent distinct cosets of Γ in Γ_i, and have zeros of high order at the other valuations $v_j(j \neq i)$. We contend that the above elements are linearly independent over K. Suppose we have a relation of linear dependence

$$\sum_{i,v} c_{iv}\, y_{iv} = 0.$$

Say c_{11} has maximal value in Γ, i.e. $v(c_{11}) \geq v(c_{iv})$ all i, v. Divide the equation by c_{11}. Then we may assume that $c_{11} = 1$, and that $v(c_{iv}) \leq 1$. Consider the value of our sum taken at v_1. All terms $y_{11}, c_{12}y_{12}, \ldots, c_{1e_1}y_{1e_1}$ have distinct values because the y's represent distinct cosets. Hence

$$v_1(y_{11} + \ldots + c_{1e_1}y_{1e_1}) \geq v_1(y_{11})$$

by Property 6 of Chapter I, § 1. On the other hand, the other terms in our sum have a very small value at v_1 by hypothesis. Hence again by that property, we have a contradiction, which proves the corollary.

2. The Riemann-Roch theorem

For the rest of this section we consider a complete normal curve V over an algebraically closed field k. The points $P, Q, \ldots$, of V are all assumed rational over k. Since two normal models of a function field K of V over k are biholomorphic, we could as well deal with the set of all local rings of K containing k. One sees trivially that the set of all such local rings coincides with the set

ALGEBRAIC GEOMETRY

of all local rings belonging to points of V, and that they are in $1-1$
correspondence with the points of V, and the k-valued places of
K over k, each such place having one of the local rings as valuation
ring. A point P of V will therefore sometimes be referred to as a
point of K.

Our main purpose is to investigate more deeply the dimension
$l(\mathfrak{a})$ of the vector space $L(\mathfrak{a})$ associated with a divisor $\mathfrak{a}$ of the
curve (we could say of the function field).

Let P be a point of V, and $\mathfrak{o}$ its local ring in K. Let $\mathfrak{p}$ be its
maximal ideal. Since k is algebraically closed, $\mathfrak{o}/\mathfrak{p}$ is canonically
isomorphic to k. We know that $\mathfrak{o}$ is a valuation ring, belonging to a
discrete valuation. Let t be a generator of the maximal ideal.
Let x be an element of $\mathfrak{o}$. Then for some constant a_0 in k, we can
write $x \equiv a_0 \bmod \mathfrak{p}$. The function $x - a_0$ is in $\mathfrak{p}$, and has a zero
at the place corresponding to P. We can therefore write $x - a_0 =
ty_0$, where y_0 is in $\mathfrak{o}$. Again by a similar argument we get
$y_0 = a_1 + ty_1$ with $y_1 \in \mathfrak{o}$, and $x = a_0 + a_1 t + y_1 t^2$. Continuing
this procedure, we obtain an expansion of x into a power series,

$$x = a_0 + a_1 t + a_2 t^2 + \dots.$$

Of course, we have repeated what was done for several variables
in Chapter VIII. In the present case, we don't need the results of
Chapter VIII. It is trivial that if each coefficient a_i is equal to 0,
then $x = 0$. It was less trivial for several variables (Krull's
theorem).

The quotient field K of $\mathfrak{o}$ can be embedded in the power series
field $k((t))$ as follows. If x is in K, then for some power t^s, the
function $t^s x$ is a unit in $\mathfrak{o}$ and has an expansion as above. This
means that x can be written

$$x = \frac{a_{-s}}{t^s} + \dots + \frac{a_{-1}}{t} + a_0 + a_1 t + \dots.$$

If u is another generator of $\mathfrak{p}$, then clearly $k((t)) = k((u))$, and
our power series field depends only on P. We denote it by K_P.
An element ξ_P of K_P can be written $\xi_P = \sum_{\nu=m}^{\infty} a_\nu t^\nu$ with $a_m \neq 0$.
If $m < 0$, we say that ξ_P has a pole of order $-m$. If $m > 0$ we
say that ξ_P has a zero of order m, and we let $m = \operatorname{ord}_P \xi_P$.

Let R^* be the cartesian product of all K_P, taken over all points P. An element of R^* can be viewed as an infinite vector $\xi = (\ldots, \xi_P, \ldots)$ where ξ_P is an element of K_P. The selection of such an element in R^* means that a random set of power series has been selected at each point P. Under componentwise addition and multiplication, R^* is a ring. It is too big for our purposes, and we shall work with the subring R consisting of all vectors such that ξ_P has no pole at P for all but a finite number of P. This ring R will be called the ring of *repartitions*. Note that our function field K is embedded in R under the mapping

$$x \to (\ldots, x, x, x, \ldots)$$

i.e. at the P-component we take x viewed as a power series in K_P. In particular, the constant field k is also embedded in R, which can be viewed as an algebra over k (infinite dimensional).

Let $\mathfrak{a}$ be a divisor on our curve. We shall denote by $\Lambda(\mathfrak{a})$ the subset of R consisting of all repartitions ξ such that $\mathrm{ord}_P \xi_P \geq -\mathrm{ord}_P \mathfrak{a}$. (If $\mathfrak{a}$ is written $\mathfrak{a} = \sum n_i P_i$, we call n_i the order of $\mathfrak{a}$ at P_i, and denote it by $\mathrm{ord}_{P_i} \mathfrak{a}$.) Then $\Lambda(\mathfrak{a})$ is immediately seen to be a k-subspace of R. The set of all such $\Lambda(\mathfrak{a})$ can be taken as a fundamental system of neighborhoods of 0 in R, and define a topology in R which thereby becomes a topological ring.

The set of functions x such that $(x) \geq -\mathfrak{a}$ is our old vector space $L(\mathfrak{a})$, and is immediately seen to be equal to $\Lambda(\mathfrak{a}) \cap K$.

If $\mathfrak{a}$ is a divisor, $\mathfrak{a} = \sum n_i P_i$, then $\sum n_i$ is called its *degree*. The purpose of this chapter is to show that $\deg(\mathfrak{a})$ and $l(\mathfrak{a})$ have the same order of magnitude, and to get precise information on $l(\mathfrak{a}) - \deg(\mathfrak{a})$. We shall eventually prove that there is a constant g depending on our field K alone such that

$$l(\mathfrak{a}) = \deg(\mathfrak{a}) + 1 - g + \delta(\mathfrak{a}),$$

where $\delta(\mathfrak{a})$ is a non-negative integer, which is 0 if $\deg(\mathfrak{a})$ is sufficiently large ($> 2g - 2$).

We now state a few trivial formulas on which we base further computations later. If A and B are two k-subspaces of R, and $A \supset B$, then we denote by $(A : B)$ the dimension of the factor space A mod B over k.

PROPOSITION 3. *Let $\mathfrak{a}$ and $\mathfrak{b}$ be two divisors. Then $\Lambda(\mathfrak{a}) \supset \Lambda(\mathfrak{b})$ if and only if $\mathfrak{a} \geq \mathfrak{b}$. If this is the case, then*

1. $(\Lambda(\mathfrak{a}) : \Lambda(\mathfrak{b})) = \deg(\mathfrak{a}) - \deg(\mathfrak{b})$, *and*
2. $(\Lambda(\mathfrak{a}) : \Lambda(\mathfrak{b})) = ((\Lambda(\mathfrak{a}) + K) : (\Lambda(\mathfrak{b}) + K)) + ((\Lambda(\mathfrak{a}) \cap K) : (\Lambda(\mathfrak{b}) \cap K)).$

Proof: The first assertion is trivial. Formula 1 is easy to prove as follows. If a point P appears in $\mathfrak{a}$ with multiplicity d and in $\mathfrak{b}$ with multiplicity e, then $d \geq e$. If t is an element of order 1 at P in K_P then the index $(t^{-d}K_P : t^{-e}K_P)$ is obviously equal to $d - e$. The index in formula 1 is clearly the sum of the finite number of local indices of the above type, as P ranges over all points in $\mathfrak{a}$ or $\mathfrak{b}$. This proves formula 1. As to formula 2, it is an immediate consequence of the elementary homomorphism theorems for vector spaces, and its formal proof will be left as an exercise to the reader.

From Proposition 3 we get a fundamental formula:

(1) $\deg(\mathfrak{a}) - \deg(\mathfrak{b}) = (\Lambda(\mathfrak{a}) + K : \Lambda(\mathfrak{b}) + K) + l(\mathfrak{a}) - l(\mathfrak{b})$

for two divisors $\mathfrak{a}$ and $\mathfrak{b}$ such that $\mathfrak{a} \geq \mathfrak{b}$. For the moment we cannot yet separate the middle index into two functions of $\mathfrak{a}$ and $\mathfrak{b}$, because we do not know that $(R : \Lambda(\mathfrak{b}) + K)$ is finite. This will be proved later.

Let y be a non-constant function in K. Let $\mathfrak{c}$ be the divisor of its poles, and write $\mathfrak{c} = \sum e_i P_i$. The points P_i in $\mathfrak{c}$ all induce the same point Q of the rational curve having function field $k(y)$, namely the point corresponding to the pole of y in $k(y)$, and the e_i are by definition the ramification indices of the discrete value group in $k(y)$ associated with the point Q, and the extensions of this value group to K. These extensions correspond to the points P_i. We shall now prove that the degree $\sum e_i$ of $\mathfrak{c}$ is equal to $[K : k(y)]$. We denote $[K : k(y)]$ by n.

Let $z_1, \ldots, z_n$ be a linear basis of K over $k(y)$. After multiplying each z_j with a suitable polynomial in $k[y]$ we may assume that they are integral over $k[y]$, i.e. that no place of K which is finite on $k[y]$ is a pole of any z_j. All the poles of the z_j are therefore among the P_i above appearing in $\mathfrak{c}$. Hence there is an integer μ_0 such that $z_j \in L(\mu_0 c)$. Let μ be a large positive integer. For any integer s satisfying $0 \leq s \leq \mu - \mu_0$ we get therefore

$$y^s z_j \in L(\mu c),$$

and so $l(\mu) \geq (\mu - \mu_0 + 1)n$.

Let N_μ be the integer $(\Lambda(\mu c) + K : \Lambda(0) + K)$, so $N_\mu \geq 0$. Putting $\mathfrak{b} = 0$ and $\mathfrak{a} = \mu c$ in the fundamental formula (1), we get

$$\mu(\textstyle\sum e_i) = N_\mu + l(\mu c) - 1$$

$$(2) \qquad\qquad \geq N_\mu + (\mu - \mu_0 + 1)n - 1.$$

Dividing (2) by μ and letting μ tend to infinity, we get $\sum e_i \geq n$. Taking into account the corollary to Theorem 1 of § 1, we get

THEOREM 2. *Let K be the function field of a curve, and $y \in K$ a non-constant function. If c is the divisor of poles of y, then* $\deg(c) = [K : k(y)]$. *Hence the degree of a divisor of a function is equal to 0 (a function has as many zeros as poles).*

Proof: if we let c' be the divisor of zeros of y then c' is the divisor of poles of $1/y$, and $[K : k(1/y)] = n$ also.

COROLLARY. $\deg(\mathfrak{a})$ *is a function of the linear equivalence class of $\mathfrak{a}$.*

A function depending only on linear equivalence will be called a class function. We see that the degree is a class function. Returning to (2), we can now write

$$\mu n \geq N_\mu + \mu n - \mu_0 n + n - 1$$

whence

$$N_\mu \leq \mu_0 n - n + 1$$

and this proves that N_μ is uniformly bounded. Hence for large μ, $N_\mu = (\Lambda(\mu c) + K : \Lambda(0) + K)$ is constant, because it is always a positive integer.

Now define a new function of divisors, $r(\mathfrak{a}) = \deg(\mathfrak{a}) - l(\mathfrak{a})$. Both $\deg(\mathfrak{a})$ and $l(\mathfrak{a})$ are class functions, the former by Theorem 2 and the latter because the map $z \to yz$ for $z \in L(\mathfrak{a})$ is a k-isomorphism between $L(\mathfrak{a})$ and $L(\mathfrak{a} - (y))$.

The fundamental formula (1) can be rewritten

(3) $0 \leq (\varLambda(\mathfrak{a}) + K : \varLambda(\mathfrak{b}) + K) = r(\mathfrak{a}) - r(\mathfrak{b})$

for two divisors $\mathfrak{a}$ and $\mathfrak{b}$ such that $\mathfrak{a} \geq \mathfrak{b}$. Put $\mathfrak{b} = 0$ and $\mathfrak{a} = \mu\mathfrak{c}$.

$$(\varLambda(\mu\mathfrak{c}) + K : \varLambda(0) + K) = r(\mu\mathfrak{c}) - r(0).$$

This and the result of the preceding paragraph show that $r(\mu\mathfrak{c})$ is uniformly bounded for all large μ.

Let now $\mathfrak{b}$ be any divisor. Take a function $z \in k[y]$ having high zeros at all points of $\mathfrak{b}$ except at those in common with $\mathfrak{c}$ (i.e. poles of y). Then for some μ, $(z) + \mu\mathfrak{c} \geq \mathfrak{b}$. Putting $\mathfrak{a} = \mu\mathfrak{c}$ in (3) above, and using the fact that $r(\mathfrak{a})$ is a class function, we get

$$r(\mathfrak{b}) \leq r(\mu\mathfrak{c})$$

and this proves that for an arbitrary divisor $\mathfrak{b}$ the integer $r(\mathfrak{b})$ is bounded. (The whole thing is of course pure magic.) This already shows that deg $(\mathfrak{b})$ and $l(\mathfrak{b})$ have the same order of magnitude. We return to this question later. For the moment, note that if we now keep $\mathfrak{b}$ fixed, and let $\mathfrak{a}$ vary in (3), then $\varLambda(\mathfrak{a})$ can be increased so as to include any element of R. On the other hand, the index in that formula is bounded because we have just seen that $r(\mathfrak{a})$ is bounded. Hence for some divisor $\mathfrak{a}$ it reaches its maximum, and for this divisor $\mathfrak{a}$ we must have $R = \varLambda(\mathfrak{a}) + K$. We state this as a theorem.

THEOREM 3. *There exists a divisor* $\mathfrak{a}$ *such that* $R = \varLambda(\mathfrak{a}) + K$. *This means that the elements of K can be viewed as a lattice in R, and that there is a neighborhood $\varLambda(\mathfrak{a})$ which when translated along all points of this lattice covers R.*

This result allows us to split the index in (1). We denote the dimension of $(R : \varLambda(\mathfrak{a}) + K)$ by $\delta(\mathfrak{a})$. We have just proved that it is finite, and (1) becomes

(4) $\mathrm{deg}\ (\mathfrak{a}) - \mathrm{deg}\ (\mathfrak{b}) = \delta(\mathfrak{b}) - \delta(\mathfrak{a}) + l(\mathfrak{a}) - l(\mathfrak{b})$

or in other words

(5) $l(\mathfrak{a}) - \mathrm{deg}\ (\mathfrak{a}) - \delta(\mathfrak{a}) = l(\mathfrak{b}) - \mathrm{deg}\ (\mathfrak{b}) - \delta(\mathfrak{b}).$

This holds for $\mathfrak{a} \geqq \mathfrak{b}$. However, since two divisors have a sup, (5) holds for any two divisors $\mathfrak{a}$ and $\mathfrak{b}$. The *genus* of K is defined to be that integer g such that

$$l(\mathfrak{a}) - \deg(\mathfrak{a}) - \delta(\mathfrak{a}) = 1 - g.$$

It is an invariant of K. Putting $\mathfrak{a} = 0$ in this definition, we see that $g = \delta(0)$, and hence that g is an integer $\geqq 0$, $g = (R : \varLambda(0) + K)$. Summarizing, we have

THEOREM 4. *There exists an integer $g \geqq 0$ depending only on K such that for any divisor $\mathfrak{a}$ we have*

$$l(\mathfrak{a}) = \deg(\mathfrak{a}) + 1 - g + \delta(\mathfrak{a}),$$

where $\delta(\mathfrak{a}) \geqq 0$.

By a *differential* λ of K we shall mean a k-linear functional of R which vanishes on some $\varLambda(\mathfrak{a})$, and also vanishes on K (considered to be embedded in R). The first condition means that λ is required to be continuous, when we take the discrete topology on k. Having proved that $(R : \varLambda(\mathfrak{a}) + K)$ is finite, we see that a differential vanishing on $\varLambda(\mathfrak{a})$ can be viewed as a functional on the factor space $R \bmod \varLambda(\mathfrak{a}) + K$, and that the set of such differentials is the dual space of our factor space, its dimension over k being therefore $\delta(\mathfrak{a})$.

Note in addition that the differentials form a vector space over K. Indeed, if λ is a differential vanishing on $\varLambda(\mathfrak{a})$, if ξ is an element of R, and y an element of K, we can define $y\lambda$ by $(y\lambda)(\xi) = \lambda(y\xi)$. The functional $y\lambda$ is again a differential, for it clearly vanishes on K, and in addition, it vanishes on $\varLambda(\mathfrak{a} + (y))$.

We shall call the sets $\varLambda(\mathfrak{a})$ *parallelotopes*. We then have the following theorem.

THEOREM 5. *If ξ is a differential, there is a maximal parallelotope $\varLambda(\mathfrak{a})$ on which λ vanishes.*

Proof: If λ vanishes on $\varLambda(\mathfrak{a}_1)$ and $\varLambda(\mathfrak{a}_2)$, and if we put $\mathfrak{a} = \sup(\mathfrak{a}_1, \mathfrak{a}_2)$, then λ vanishes on $\varLambda(\mathfrak{a})$. Hence to prove our theorem it will suffice to prove that the degree of $\mathfrak{a}$ is bounded. Let $\mathfrak{b}$ be an arbitrary divisor. If $y \in L(\mathfrak{b})$, so $(y) \geqq -\mathfrak{b}$, then $y\lambda$ vanishes on

$\Lambda(\mathfrak{a} + (y))$ which contains $\Lambda(\mathfrak{a} - \mathfrak{b})$ because $\mathfrak{a} + (y) \geqq \mathfrak{a} - \mathfrak{b}$. If $y_1, \ldots, y_n$ are linearly independent over k, then so are $y_1\lambda, \ldots, y_n\lambda$. Hence we get

$$\delta(\mathfrak{a} - \mathfrak{b}) \geqq l(\mathfrak{b}).$$

Using Theorem 4, we get

$$l(\mathfrak{a} - \mathfrak{b}) - \deg (\mathfrak{a}) + \deg (\mathfrak{b}) - 1 + g \geqq l(\mathfrak{b})$$
$$\geqq \deg (\mathfrak{b}) + 1 - g + \delta(\mathfrak{b}),$$

and from this we see that $\deg (\mathfrak{a}) \leqq l(\mathfrak{a} - \mathfrak{b}) + 2g - 2 - \delta(\mathfrak{b})$. Take $\mathfrak{b}$ to be a positive divisor of very large degree. Then $L(\mathfrak{a} - \mathfrak{b}) = 0$ because a non-zero function cannot have more zeros than poles. Since $\delta(\mathfrak{b}) \geqq 0$, we see that $\deg (\mathfrak{a}) \leqq 2g - 2$ and is therefore bounded. This proves our theorem.

THEOREM 6. *The differentials form a 1-dimensional K-space.*

Proof: Suppose we have two differentials λ and μ which are linearly independent over K. Suppose $x_1, \ldots, x_n$ and $y_1, \ldots, y_n$ are two sets of elements of K which are linearly independent over k. Then the differentials $x_1\lambda, \ldots, x_n\lambda, y_1\mu, \ldots, y_n\mu$ are linearly independent over k, for otherwise we would have a relation

$$\sum a_i x_i \lambda + \sum b_i y_i \mu = 0.$$

Letting $x = \sum a_i x_i$ and $y = \sum b_i y_i$, we get $x\lambda + y\mu = 0$, contradicting the independence of λ, μ over K.

Both λ and μ vanish on some parallelotope $\Lambda(\mathfrak{a})$, for if λ vanishes on $\Lambda(\mathfrak{a}_1)$ and μ vanishes on $\Lambda(\mathfrak{a}_2)$, we put $\mathfrak{a} = \inf (\mathfrak{a}_1, \mathfrak{a}_2)$, and $\Lambda(\mathfrak{a}) = \Lambda(\mathfrak{a}_1) \cap \Lambda(\mathfrak{a}_2)$. Let $\mathfrak{b}$ be an arbitrary divisor. If $y \in L(\mathfrak{b})$, so that $(y) \geqq - \mathfrak{b}$, then $y\lambda$ vanishes on $\Lambda(\mathfrak{a} + (y))$ which contains $\Lambda(\mathfrak{a} - \mathfrak{b})$ because $\mathfrak{a} + (y) \geqq \mathfrak{a} - \mathfrak{b}$. Similarly, $y\mu$ vanishes on $\Lambda(\mathfrak{a} - \mathfrak{b})$, and by definition and the remark at the beginning of our proof, we conclude that

$$\delta(\mathfrak{a} - \mathfrak{b}) \geqq 2l(\mathfrak{b}).$$

Using Theorem 4, we get

$$l(\mathfrak{a} - \mathfrak{b}) - \deg (\mathfrak{a}) + \deg (\mathfrak{b}) - 1 + g \geqq 2l(\mathfrak{b})$$
$$\geqq 2 (\deg (\mathfrak{b}) + 1 - g + \delta(\mathfrak{b}))$$
$$\geqq 2 \deg (\mathfrak{b}) + 2 - 2g.$$

If we take $\mathfrak{b}$ to be a positive divisor of very large degree, then $L(\mathfrak{a} - \mathfrak{b})$ consists of 0 alone, because a function cannot have more zeros than poles. Since deg $(\mathfrak{a})$ is constant in the above inequality, we get a contradiction, and thereby prove the theorem.

If λ is a non-zero differential, then all differentials are of type $y\lambda$. If $\Lambda(\mathfrak{a})$ is the maximal parallelotope on which λ vanishes, then clearly $\Lambda(\mathfrak{a} + (y))$ is the maximal parallelotope on which $y\lambda$ vanishes. We get therefore a linear equivalence class of divisors: if we define the divisor (λ) associated with λ to be $\mathfrak{a}$, then the divisor associated with $y\lambda$ is $\mathfrak{a} + (y)$. This divisor class is called the *canonical class* of K, and a divisor in it is called a *canonical divisor*.

Theorem 6 allows us to complete Theorem 4 by giving more information on $\delta(\mathfrak{a})$; we can now state the complete Riemann-Roch theorem.

THEOREM 7. *Let $\mathfrak{a}$ be an arbitrary divisor of K. Then*

$$l(\mathfrak{a}) = \deg(\mathfrak{a}) + 1 - g + l(\mathfrak{c} - \mathfrak{a}),$$

where $\mathfrak{c}$ is any divisor of the canonical class. In other words, $\delta(\mathfrak{a}) = l(\mathfrak{c} - \mathfrak{a})$.

Proof: Let $\mathfrak{c}$ be the divisor which is such that $\Lambda(\mathfrak{c})$ is the maximal parallelotope on which a non-zero differential λ vanishes. If $\mathfrak{b}$ is an arbitrary divisor and $y \in L(\mathfrak{b})$, then we know that $y\lambda$ vanishes on $\Lambda(\mathfrak{c} - \mathfrak{b})$. Conversely, by Theorem 6, any differential vanishing on $\Lambda(\mathfrak{c} - \mathfrak{b})$ is of type $z\lambda$ for some $z \in K$, and the maximal parallelotope on which $z\lambda$ vanishes is $(z) + \mathfrak{c}$, which must therefore contain $\Lambda(\mathfrak{c} - \mathfrak{b})$. This implies that $(z) \geqq - \mathfrak{b}$, i.e. $z \in L(\mathfrak{b})$. We have therefore proved that $\delta(\mathfrak{c} - \mathfrak{b})$ is equal to $l(\mathfrak{b})$. The divisor $\mathfrak{b}$ was arbitrary, and hence we can replace it by $\mathfrak{c} - \mathfrak{a}$, thereby proving our theorem.

COROLLARY 1. *If $\mathfrak{c}$ is a canonical divisor, then $l(\mathfrak{c}) = g$.*

Proof: Put $\mathfrak{a} = 0$ in the Riemann-Roch theorem. Then $L(\mathfrak{a})$ consists of the constants alone, and so $l(\mathfrak{a}) = 1$. Since deg $(0) = 0$, we get what we want.

COROLLARY 2. *The degree of the canonical class is $2g - 2$.*

242   ALGEBRAIC GEOMETRY

Proof: Put $\mathfrak{a} = \mathfrak{c}$ in the Riemann-Roch theorem, and use Corollary 1.

COROLLARY 3. If deg $(\mathfrak{a}) > 2g - 2$, then $\delta(\mathfrak{a}) = 0$.

Proof: $\delta(\mathfrak{a})$ is equal to $l(\mathfrak{c} - \mathfrak{a})$. Since a function cannot have more zeros than poles, $L(\mathfrak{c} - \mathfrak{a}) = 0$ if deg $(\mathfrak{a}) > 2g - 2$.

3. Residues in power series fields

The results of this section will be used as lemmas to prove that the sum of the residues of a differential form in a function field of dimension 1 is 0.

Let $k((t))$ be a power series field, the field of coefficients being arbitrary (not necessarily algebraically closed). If u is an element of that field which can be written $u = a_1 t + a_2 t^2 + \ldots$ with $a_1 \neq 0$, then it is clear that $k((u)) = k((t))$. The order of an element of $k((t))$ can be computed in terms of u or of t. We call an element of order 1 a *local parameter* of $k((t))$.

Our power series field admits a derivation D_t defined in the obvious manner. Indeed, if $y = \sum a_\nu t^\nu$ is an element of $k((t))$ one verifies immediately that $D_t y = \sum \nu a_\nu t^{\nu-1}$ is a derivation. We sometimes denote $D_t y$ by dy/dt. There is also a derivation $D_u y$ defined in the same manner, and the classical chain rule $D_u y \cdot D_t u = D_t y$ (or better $dy/du \cdot du/dt = dy/dt$) holds here because it is a formal result.

If $y = \sum a_\nu t^\nu$, then a_{-1} (the coefficient of t^{-1}) is called the *residue of y with respect to t*, and denoted by $\mathrm{res}_t(y)$.

PROPOSITION 4. *Let x and y be two elements of $k((t))$, and let u be another parameter of $k((t))$. Then*

$$\mathrm{res}_u\left(y\frac{dx}{du}\right) = \mathrm{res}_t\left(y\frac{dx}{dt}\right).$$

Proof: It clearly suffices to show that for any element y of $k((t))$ we have $\mathrm{res}_t(v) = \mathrm{res}_u(y\, dt/du)$. Since the residue is k-linear as a function of power series, and vanishes on power series which have a zero of high order, it suffices to prove our proposition for $y = t^n$ (n being an integer). Furthermore, our result is obviously true under the trivial change of parameter $t = au$, where a is a non-zero constant. Hence we may assume $t = u + a_2 u^2 + \ldots$, and $dt/du = 1 + 2a_2 u + \ldots$. We have to show that $\operatorname{res}_u(t^n dt/du) = 1$ when $n = -1$, and 0 otherwise.

When $n \geq 0$, the proposition is obvious, because $t^n dt/du$ contains no negative powers of t.

When $n = -1$, we have

$$\frac{1}{t} \frac{dt}{du} = \frac{1 + 2a_2 u + \ldots}{u + 2a_2 u^2 + \ldots} = \frac{1}{u} + \ldots,$$

and hence the residue is equal to 1, as desired.

When $n < -1$, we consider first the case in which the characteristic is 0. In this case, we have

$$\operatorname{res}_u(t^n dt/du) = \operatorname{res}_u \left(\frac{d}{du} \left(\frac{1}{n+1} t^{n+1} \right) \right)$$

and this is 0 for $n \neq -1$.

For arbitrary characteristic, and fixed $n < -1$, we have for $m > 1$,

$$\frac{1}{t^m} \frac{dt}{du} = \frac{1 + 2a_2 u + \ldots}{u^m(1 + a_3 u + \ldots)}$$

$$= \frac{F(a_2, a_3, \ldots)}{t} + \ldots,$$

where $F(a_2, a_3, \ldots)$ is a formal polynomial with integer coefficients. It is the same for all fields, and contains only a finite number of the coefficients a_i. Hence the truth of our proposition is a formal consequence of the result in characteristic 0, because we have just seen that in that case, our polynomial $F(a_2, a_3, \ldots)$ is identically 0. This proves the proposition.

In view of Proposition 4, we shall call an expression of type ydx (with x and y in the power series field) a *differential form of that field*, and the *residue* res (ydx) of that differential form is defined to be the residue $\operatorname{res}_t(y\, dx/dt)$ taken with respect to any parameter t of our field.

We shall need a more general formula than that of Proposition 4.

Given a power series field $k((u))$, let t be a non-zero element of that field of order $m \geq 1$. After multiplying t by a constant if necessary, we can write

$$t = u^m + b_1 u^{m+1} + b_2 u^{m+2} + \ldots$$
$$= u^m(1 + b_1 u + b_2 u^2 + \ldots).$$

Then the power series field $k((t))$ is contained in $k((u))$. In fact, one sees immediately that the degree of $k((u))$ over $k((t))$ is exactly equal to m. Indeed, by recursion, one can express any element y of $k((u))$ in the following manner

$$y = f_0(t) + f_1(t)u + \ldots + f_{m-1}(t)u^{m-1},$$

with $f_i(t) \in k((t))$. Furthermore, the elements $1, u, \ldots, u^{m-1}$ are linearly independent over $k((t))$, because our power series field $k((u))$ has a discrete valuation where u is an element of order 1, and t has order m. If we had a relation as above with $y = 0$, then two terms $f_i(t)u^i$ and $f_j(t)u^j$ would necessarily have the same absolute value with $i \neq j$. This obviously cannot be the case. Hence the degree of $k((u))$ over $k((t))$ is equal to m, and is equal to the ramification index of the valuation in $k((t))$ having t as an element of order 1 with respect to the valuation in $k((u))$ having u as element of order 1.

The following proposition gives the relations between the residues taken in $k((u))$ or in $k((t))$. By Tr we shall denote the trace from $k((u))$ to $k((t))$.

PROPOSITION 5. *Let $k((u))$ be a power series field, and let t be a non-zero element of that field, of order $m \geq 1$. Let y be an element of $k((u))$. Then*

$$\operatorname{res}_u \left(y \frac{dt}{du} du \right) = \operatorname{res}_t (Tr(y)dt).$$

Proof: We have seen that the powers $1, u, \ldots, u^{m-1}$ form a basis for $k((u))$ over $k((t))$, and the trace of an element y of $k((u))$ can be computed from the matrix representing y on this basis. Multiplying t by a non-zero constant does not change the validity of the proposition. Hence we may assume that

$t = u^m + b_1 u^{m+1} + \ldots = u^m(1 + b_1 u + b_2 u^2 + \ldots)$, with $b_\nu \in k$. One can solve recursively

$$u^m = f_0(t) + f_1(t)u + \ldots + f_{m-1}(t)u^{m-1},$$

where $f_i(t)$ are elements of $k((t))$, and the coefficients of $f_i(t)$ are universal polynomials in $b_1, b_2, \ldots$, with *integer* coefficients, i.e. each $f_i(t)$ can be written

$$f_i(t) = \sum P_{i\nu}(b)t^\nu,$$

where each $P_{i\nu}(b)$ is a polynomial with integer coefficients, involving only a finite number of b's.

The matrix representing an arbitrary element

$$g_0(t) + g_1(t)u + \ldots + g_{m-1}(t)u^{m-1}$$

of $k((u))$ is therefore of type

$$\begin{pmatrix} G_{0,1}(t) & \cdots & G_{0,\,m-1}(t) \\ \cdots\cdots\cdots\cdots\cdots \\ G_{m-1,\,0}(t) & \cdots & G_{m-1,\,m-1}(t) \end{pmatrix}$$

where $G_{\nu\mu}(t) \in k((t))$, and where the coefficients of the $G_{\nu\mu}(t)$ are universal polynomials with integer coefficients in the b's and in the coefficients of the $g_i(t)$. This means that our formula, if it is true, is a formal identity having nothing to do with characteristic p, and that our verification can be carried out in characteristic 0.

This being the case, we can write $t = v^m$, where $v = u + c_2 u^2 + \ldots$. is another parameter of the field $k((u))$. This can be done by taking the binomial expansion for $(1 + b_1 u + \ldots)^{1/m}$. In view of Proposition 4, it will suffice to prove that

$$\operatorname{res}_v\left(y \frac{dt}{dv} \, dv\right) = \operatorname{res}_t(Tr(y)dt).$$

By linearity, it suffices to prove this for $y = v^j$, $-\infty < j < +\infty$. (If y has a very high order, then both sides are obviously equal to 0, and y can be written as a sum involving a finite number of terms $a_j v^j$, and an element of very high order.)

If we write $j = ms + r$ with $0 \leq r \leq m - 1$, then $v^j = t^s v^r$ and $Tr(v^j) = t^s Tr(v^r)$. We have trivially

$$Tr(v^r) = \begin{cases} 0 & \text{if } r \neq 0 \\ m & \text{if } r = 0 \end{cases} \quad \text{whence} \quad Tr(v^j) = \begin{cases} mt^s & \text{if } j = ms \\ 0 & \text{otherwise.} \end{cases}$$

Consequently, we get

$$\operatorname{res}_t(Tr(v^j)dt) = \begin{cases} m & \text{if } j = -m \\ 0 & \text{otherwise.} \end{cases}$$

On the other hand, our first expression in terms of v is equal to $\operatorname{res}_v(v^j m v^{m-1} dv)$, which is obviously equal to what we just obtained for the right-hand side. This proves our proposition.

In the preceding discussion, we started with a power series field $k((u))$ and a subfield $k((t))$. We conclude this section by showing that this situation is typical of power series field extensions.

Let $F = k((t))$ be a given power series field over an algebraically closed field k. We have a canonical k-valued place of F, mapping t on 0. Let E be a finite algebraic extension of F. Then the discrete valuation of F extends in at least one way to E, and so does our place. Let u be an element of E of order 1 at the extended valuation, which is discrete by the corollary of Proposition 7, Chapter I, § 6. If e is the ramification index, then we know by this same proposition that $e \leq [E : F]$. We shall show that $e = [E : F]$ and hence that the extension of our place is unique.

An element y of E which is finite under the place has an expansion $y = a_0 + a_1 u + \ldots + a_{e-1} u^{e-1} + t y_1$, where y_1 is in E and is also finite. This comes from the fact that u^e and t have the same order in the extended valuation. Similarly, y_1 has also such an expansion, $y_1 = b_0 + b_1 u + \ldots + b_{e-1} u^{e-1} + t y_2$. Substituting this expression for y_1 above, and continuing the procedure, we see that we can write

$$y = f_0(t) + f_1(t)u + \ldots + f_{e-1}(t)u^{e-1},$$

where $f_i(t)$ is a power series in $k((t))$. Since the powers $1, u, \ldots, u^{e-1}$ are clearly linearly independent over $k((t))$, this proves that $e = [E : F]$, and that the extension of the place is unique.

Furthermore, to every element of E we can associate a power series in $\sum_{\mu=m}^{\infty} a_\mu u^\mu$, and one sees that every such power series arises from an element of E, because we can always replace u^e by ty, where y is finite under the place, and we can therefore solve recursively for a linear combination of the powers $1, u, \ldots, u^{e-1}$ with coefficients in $k((t))$. Summarizing, we get

PROPOSITION 6. *Let $k((t))$ be a power series field over an algebraically closed field k. Then the natural k-valued place of $k((t))$ has a unique extension to any finite algebraic extension of $k((t))$. If E is such an extension, and u is an element of order 1 in the extended valuation, then E may be identified with the power series field $k((u))$.*

4. The sum of the residues

We return to global considerations, and consider a function field K of dimension 1 over an algebraically closed field k. The points P of a model of K over k are identified with the k-valued places of K over k. For each such point, we have an embedding $K \to K_P$ of K into a power series field $k((t)) = K_P$ as in § 2. Our first task will be to compare the derivations in K with the derivations in $k((t))$ discussed in § 3.

THEOREM 8. *Let y be an element of K. Let $t \in K$ be a local parameter at the point P, and let z be the element of K which is such that $dy = zdt$. If dy/dt is the derivative of y with respect to t taken formally from the power series expansion of y, then $z = dy/dt$.*

Proof: The statement of our theorem depends on the fact that every differential form of K can be written zdt for some z, by the results of Chapter VII. We know that K is separable algebraic over $k(t)$, and the irreducible polynomial equation $f(t, y) = 0$ of y over $k(t)$ is such that $f_y(t, y) \neq 0$. On the one hand we have

$$0 = f_t(t, y)dt + f_y(t, y)dy,$$

whence $z = -f_t(t, y)/f_y(t, y)$ (cf. Proposition 2 of Chapter VII,

§2) and on the other hand, if we differentiate with respect to t the relation $f(t, y) = 0$ in the power series field, we get

$$0 = f_t(t, y) + f_v(t, y) \frac{dy}{dt}.$$

This proves our theorem.

Let ω be a differential form of K. Let P be a point of K, and t a local parameter, selected in K. Then we can write $\omega = y dt$ for some $y \epsilon K$. Referring to Proposition 4 of § 3, we can define the *residue of ω at P* to be the residue of $y dt$ at t, i.e.

$$\operatorname{res}_P(\omega) = \operatorname{res}_t(y).$$

If ω is written $x dz$ for $x, z \epsilon K$, this residue is also written $\operatorname{res}_P(x dz)$. We now state the main theorem of this section.

THEOREM 9. *Let K be the function field of a curve over an algebraically closed constant field k. Let ω be a differential form of K. Then*

$$\sum_P \operatorname{res}_P(\omega) = 0.$$

(The sum is taken over all points P, but is actually a finite sum since the differential form has only a finite number of poles.)

Proof: The proof is carried out in two steps, first in a rational function field, and then in an arbitrary function field using Proposition 5.

Consider first the case where $K = k(x)$, where x is a single transcendental quantity over k. The points P are in $1 - 1$ correspondence with the specializations of x in k, and with the specialization $1/x \to 0$ (i.e. the place sending $x \to \infty$). If P is not the point sending x to infinity, but, say the point $x = a$, $a \epsilon k$, then $x - a$ can be selected as parameter at P, and the residue of a differential form $y dx$ is the residue of y in its expansion in terms of $x - a$. The situation is the same as in complex variables.

We expand y into partial fractions,

$$y = \sum c_{\mu i} / (x - b_\mu)^i + f(x)$$

where $f(x)$ is a polynomial in $k[x]$. To get $\operatorname{res}_P(y dx)$ we need consider only the coefficient of $(x - a)^{-1}$ and hence the sum of the residues taken over all P finite on x is equal to $\sum_\mu c_{\mu 1}$.

Now suppose P is the point at infinity. Then $t = 1/x$ is a local parameter, and $dx = -1/t^2\,dt$. We must find the coefficient of $1/t$ in the expression $-y1/t^2$. It is clear that the residue at t of $-1/t^2 f(1/t)$ is equal to 0. The other expression can be expanded as follows:

$$-\frac{1}{t^2} \cdot \sum_{\mu, i} \frac{c_{\mu i}}{\left(\dfrac{1}{t} - b_\mu\right)^i} = -\sum_{\mu, i} c_{\mu i} t^{i-2} (1 + tb_\mu + \ldots)^i$$

and from this we get a contribution to the residue only from the first term, which gives precisely $-\sum c_{\mu 1}$. This proves our theorem in the case of a purely transcendental field.

Next, suppose we have a finite separable algebraic extension K of a purely transcendental field $F = k(x)$ of dimension 1 over the algebraically closed constant field k. Let Q be a point of F, and t a local parameter at Q in F. Let P be a point of K lying above Q, and let u be a local parameter at P in K. Under the discrete valuation at P in K extending that of Q in F, we have $\operatorname{ord}_P t = e \cdot \operatorname{ord}_P u$. The power series field $k((u))$ is a finite extension of degree e of $k((t))$. Thus for each P we get an embedding of K in a finite algebraic extension of $k((t))$, and the place on K at P is induced by the canonical place of the power series field $k((u))$.

Let $P_i (i = 1, \ldots, s)$ be the points of K lying above Q. Let A be the algebraic closure of $k((t))$. The discrete valuation of $k((t))$ extends uniquely to a valuation of A, which is discrete on every subfield of A finite over $k((t))$ (Proposition 6 of § 3). Suppose $K = F(y)$ is generated by one element y, satisfying the irreducible polynomial $g(Y)$ with leading coefficient 1 over F. It splits into irreducible factors over $k((t))$, say

(1) $$g(Y) = g_1(Y) \cdots g_r(Y)$$

of degrees $d_j (j = 1, \ldots, r)$. Let y_j be a root of $g_j(Y)$. Then the mapping $y \to y_j$ induces an isomorphism of K into A. Two roots of the same g_j are conjugate over $k((t))$, and give rise to conjugate fields. By the uniqueness of the extension of the place, the induced place on K is therefore the same for two such conjugate embeddings. The ramification index relative to this embedding is d_j, and we see from (1) that $\sum d_j = n$. By Theorem 2 of § 2 we now conclude that two distinct polynomials g_j give rise to two distinct places on K, and that $s = r$. We can therefore identify the fields $k((t))(y_i)$ with the fields K_{P_i}.

For each $i = 1, \ldots, r$ denote by Tr_i the trace from the field K_{P_i} to F_Q.

PROPOSITION 7. *The notation being as above, let Tr be the trace from K to F. Then for any $y \in K$, we have*

$$Tr(y) = \sum_{i=1}^{r} Tr_i(y).$$

Proof: Suppose y is a generator of K over F. If $[K : F] = n$, then $Tr(y)$ is the coefficient of Y^{n-1} in the irreducible polynomial $g(Y)$ as above. A similar remark applies to the local traces, and our formula is then obvious from (1). If y is not a generator, let z be a generator. For some constant $c \in k$, $w = y + cz$ is a generator. The formula being true for cz and for w, and both sides of our equation being linear in y, it follows that the equation holds for y, as desired.

The next proposition reduces the theorem for an arbitrary function field K to a rational field $k(x)$.

PROPOSITION 8. *Let k be algebraically closed. Let $F = k(x)$ be a purely transcendental extension of dimension 1, and K a finite algebraic separable extension of F. Let Q be a point of F, and $P_i(i = 1, \ldots, r)$ the points of K lying above Q. Let y be an element of K, and let Tr denote the trace from K to F. Then*

$$\mathrm{res}_Q(Tr(y)dx) = \sum_{i=1}^{r} \mathrm{res}_{P_i}(ydx).$$

Proof: Let t be a local parameter at Q in $k(x)$, and let u_i be a local parameter in K at P_i. Let Tr_i denote the local trace from K_{P_i} to F_Q. We have

$$\sum \mathrm{res}_{P_i}(ydx) = \sum \mathrm{res}_{P_i}\left(y\frac{dx}{dt}dt\right)$$

$$= \sum \mathrm{res}_{P_i}\left(y\frac{dx}{dt}\frac{dt}{du_i}du_i\right)$$

and using Proposition 5 of § 3, we see that this is equal to

$$\sum \mathrm{res}_Q(Tr_i(ydx/dt).$$

Since dx/dt is an element of $k(x)$, the trace is homogeneous with respect to this element, and the above expression is equal to

$$\sum \mathrm{res}_Q\left(Tr_i(y)\frac{dx}{dt}dt\right) = \sum \mathrm{res}_Q(Tr_i(y)dx)$$

$$= \mathrm{res}_Q(\sum Tr_i(y)dx)$$

$$= \mathrm{res}_Q(Tr(y)dx)$$

thereby proving our proposition.

Theorem 9 now follows immediately, because a differential form can be written ydx, where K is separable algebraic over $k(x)$.

Our theorem will allow us to identify differential forms of a function field K with the differentials introduced in § 2, as k-linear functionals on the ring R of repartitions which vanish on some $\Lambda(\mathfrak{a})$ and on K. This is done in the following manner. Let $\xi = (\ldots, \xi_P, \ldots)$ be a repartition. Let ydx be a differential form of K. Then the map

$$\lambda : \xi \to \sum_P \mathrm{res}_P(\xi_P ydx)$$

is a k-linear map of R into k. Here of course, in the expression $\mathrm{res}_P(\xi_P ydx)$ one views y and x as elements of K_P. It is also clear that all but a finite number of terms of our sum are 0.

Our k-linear map vanishes on some $\Lambda(\mathfrak{a})$, because the differential form has only a finite number of poles. Theorem 9 shows that it vanishes on K. It is therefore a differential, and in this way we obtain an isomorphism of the K-vector space of differential forms into the K-vector space of differentials. Since both spaces have dimension 1 over K (the latter by Theorem 6 of § 2), this isomorphism is surjective.

To make our identification complete, we shall show that the divisor (λ) of the differential defined in § 2 is the same as the divisor (ydx) of the differential form introduced in Chapter VII, § 2. This is easily done. Suppose $\operatorname{ord}_P(ydx) = m_P$. If $\operatorname{ord}_P(\xi_P) \geq -m_P$, then $\operatorname{ord}_P(\xi_P ydx) \geq 0$, and the residue $\operatorname{res}_P(\xi_P ydx)$ is 0. Hence the differential λ vanishes on $\Lambda(\mathfrak{a})$, where $\mathfrak{a} = (ydx)$. On the other hand, if $\Lambda(\mathfrak{b})$ is the maximal parallelotope on which λ vanishes, then $\Lambda(\mathfrak{b}) \supset \Lambda(\mathfrak{a})$, and $\mathfrak{b} \geq \mathfrak{a}$. If $\mathfrak{b} > \mathfrak{a}$, then for some P, the coefficient of P in $\mathfrak{b}$ is $> m_P$, and hence the repartition $(\ldots 0, 0, 1/t^{m_P+1}, 0, 0, \ldots)$ lies in $\Lambda(\mathfrak{b})$. One sees immediately from the definitions that $\operatorname{res}_P(t^{-m_P-1}ydx) \neq 0$, and hence λ cannot vanish on $\Lambda(\mathfrak{b})$. Summarizing we have

THEOREM 10. *Let K be a function field of dimension* 1 *over the algebraically closed constant field k. Each differential form ydx of K gives rise to a differential*

$$\lambda : \xi \to \sum_P \operatorname{res}_P(\xi_P ydx),$$

and this induces a K-isomorphism of the K-space of differential forms onto the K-space of differentials. Furthermore, the divisors (ydx) of Chapter VII and (λ) of § 2 are equal.

5. Another proof for the sum of the residues

It is of some interest that the local result of Proposition 5 in § 3 can be avoided, provided we make use of the global results of § 2. We shall sketch a new proof due to Artin of the fact that the sum of the residues of a differential form is equal to 0.

We need the following results: The invariance of the residue with respect to local parameters; the fact that if x, y are two elements of the function field, and t is a local parameter at a point P, then the unique function z such that $zdx = dy$ is equal to $D_t y/D_t x$, where D_t is the derivation with respect to t in the local power series field; and finally the fact that if K is a finite separable extension of the rational field $k(x)$, then the global trace Tr from K to $k(x)$ is equal to the sum of the local traces Tr_i. (Here of course by the local traces we mean that if Q is any point of $k(x)$, and P_i the points of K lying above Q, then Tr_i is the trace from K_{P_i} to $k(x)_Q$.) All these results have of course been proved before.

Now if x is an element of K such that K is separable algebraic over $k(x)$, then we define a differential λ_x on R in the following manner. In the rational field, given a repartition $(\ldots,\xi_Q,\ldots)$ we let $\lambda_x(\xi) = \sum_Q \operatorname{res}_Q(\xi_Q dx)$. It is easily proved that in a rational field, the sum of the residues of a differential form ydx is 0. Hence λ_x vanishes on the functions of $k(x)$. If now $\xi = (\ldots, \xi_P,\ldots)$ is a repartition of K, we define $\lambda_x(\xi) = \sum_P \lambda_x(Tr_P(\xi_P))$, where Tr_P denotes the trace from K_P to $k(x)_Q$. Since the global trace is the sum of the local traces, it follows that λ_x again vanishes on the elements of K. It is easily seen that λ_x vanishes on a parallelotope, and is non-trivial because K is separable algebraic over $k(x)$. We have therefore constructed a non-zero differential. We shall now prove that if P is any point of K, and ξ_P an element of K_P, identified with the repartition $(\ldots, 0, 0, \xi_P, 0, 0, \ldots)$ which is 0 everywhere except at P where it is equal to ξ_P, then $\lambda_x(\xi_P) = \operatorname{res}_P(\xi_P dx)$. Since λ_x vanishes on the elements of K, it will follow that $\sum \operatorname{res}_P(ydx) = 0$, as desired.

If x is a local parameter at P, then K_P coincides with $k(x)_Q$, where Q is the point on $k(x)$ induced by P. Hence in that case our result follows from the definition of λ_x. Let now A be an arbitrary point of K, and let t be a local parameter at A in K. Let $z = dx/dt$. By definition, $(z\lambda_t)(\xi_A) = \operatorname{res}_A(z\xi_A dt)$, and by the invariance of the residue, this is equal to $\operatorname{res}_A(\xi dx)$. If we can show that $z\lambda_t$ is equal to λ_x, then the proof will be complete. Select P to be a point where both x and t are local parameters (the set of points where x or t is not a parameter is finite). Then for that P, we have

$$\operatorname{res}_P(\xi_P dx) = \operatorname{res}_P\left(\xi_P \frac{dx}{dt} dt\right),$$

and hence at that P, $z\lambda_t$ and λ_x coincide as functions on k_P (embedded in the repartitions). However, we know that there is a maximal parallelotope on which a differential vanishes. Hence if two differentials coincide at one point, they must be equal globally, and we get $z\lambda_t = \lambda_x$. This proves what we wanted.

6. Construction of the Jacobian variety

Let C be a complete non-singular variety of dimension 1. Such a variety will be called a curve. We assume that C is defined over a field k_0, and that all fields in the sequel contain k_0. We assume that the genus g of C is > 0.

LEMMA. *Let $P_1, \ldots, P_m$ be m independent generic points of C over some field k, and let K be the field fixed under the group of automorphisms of $k(P_1, \ldots, P_m)$ which leave k fixed and permute the P_i among themselves. Then K is the smallest field of rationality of the cycle $\sum P_i$ which is prime rational over K.*

Proof: By Galois theory, $k(P_1, \ldots, P_m)$ is a finite Galois extension of K. Any conjugate of, say, P_1 over K is one of the P_i, and every P_i is a conjugate. Hence our cycle is prime rational over K. Let E be the smallest field of rationality of our cycle. We must show that $EK = E$. If σ is an automorphism of the universal domain leaving E fixed, then σ leaves our cycle fixed, and hence permutes the P_i. Hence σ leaves K fixed. Therefore EK is purely inseparable over E. On the other hand, all the coefficients of our cycle being equal to 1, $E(P_1, \ldots, P_m)$ is separable algebraic over E, and all elements of K must *a fortiori* be separable algebraic over E. Thus K is contained in E.

PROPOSITION 9. *Let $P_1, \ldots, P_g$ be g independent generic points of C over a field k, and let $\mathfrak{a}$ be a divisor on C rational over k, and of degree 0. Then there is one and only one positive divisor $\mathfrak{q}$ linearly equivalent to $\sum P_i + \mathfrak{a}$. It is of type $\sum Q_i$, where the Q_i are g independent generic points over k, and the smallest field of rationality of $\mathfrak{q}$ is the same as that of $\mathfrak{p} = \sum P_i$.*

Proof: Either there exists a function f on C, defined over k such that $\mathfrak{a} = (f)$, or $l(\mathfrak{a}) = 0$. In the first case, $l(\mathfrak{a}) = 1$. By the Riemann Roch theorem, $l(\mathfrak{c} - \mathfrak{a}) = g$. Note that we can take $\mathfrak{c}$ rational over k because we can take a differential form $y dx$ in the function field $k(C)$. If P_1 is generic over k, then by general principles (cf. the first part of the proof of Theorem 2, Chapter VI, § 3), we have

$$(1) \qquad l(\mathfrak{c} - \mathfrak{a} - P_1) \leq l(\mathfrak{c} - \mathfrak{a}) \leq l(\mathfrak{c} - \mathfrak{a} - P_1) + 1.$$

Not all functions in $L(\mathfrak{c} - \mathfrak{a})$ can have a zero at P_1, because the space of such functions has a basis consisting of elements defined over k (Theorem 5 of Chapter VI, § 5). Hence the equality cannot hold on the left, and we have $l(\mathfrak{c} - \mathfrak{a} - P_1) = l(\mathfrak{c} - \mathfrak{a}) - 1$, so $l(\mathfrak{a} + P_1) = l(\mathfrak{a})$. We can repeat this type of argument inductively g times, and we get $l(\mathfrak{a} + \mathfrak{p}) = 1$. Hence there is only one positive divisor linearly equivalent to $\mathfrak{a} + \mathfrak{p}$, namely $\mathfrak{p}$ itself.

Similarly, suppose $l(\mathfrak{a}) = 0$. Then $l(\mathfrak{c} - \mathfrak{a}) = g - 1$. As before in (1), we cannot have the equality on the left, and we can proceed inductively $g - 1$ times to get $l(\mathfrak{c} - \mathfrak{a} - P_1 - \ldots - P_{g-1}) = 0$. Adding one more generic point P_g, we obtain $l(\mathfrak{c} - \mathfrak{a} - \mathfrak{p}) = 0$, whence $l(\mathfrak{a} + \mathfrak{p}) = 1$.

This proves the existence and uniqueness of our divisor $\mathfrak{q}$, which consists of g points since a function has as many zeros as poles. We do not yet know that they are distinct, and we must show that $\mathfrak{q} = \sum_{i=1}^{g} Q_i$ where the Q_i are generic independent. We can find a function φ on C defined over the smallest field of rationality for $\mathfrak{p}$ containing k, such that $(\varphi) = \mathfrak{q} - \mathfrak{a} - \mathfrak{p}$ by Corollary 2 of Theorem 5, Chapter VI, § 5. This shows that $\mathfrak{q}$ is rational over that field. On the other hand, since $\mathfrak{q}$ and $\mathfrak{a} + \mathfrak{p}$ are linearly equivalent, we get $l(\mathfrak{q}) = 1$. By symmetry, we see that $\mathfrak{q}$ and $\mathfrak{p}$ have the same smallest field of rationality, which has transcendence degree g over k. This implies that the points Q_i must be generic independent and concludes our proof.

We shall now suppose that C has a positive divisor $\mathfrak{o}$ of degree g rational over k. Let $\mathfrak{p} = \sum P_i$ be a generic divisor of degree g as before, and denote by $k(\mathfrak{p})$ its smallest field of rationality. Let U be any model of the field $k(\mathfrak{p})$ over k. We are going to define a normal law of composition on U by means of the Riemann-Roch theorem. Let $\mathfrak{p}, \mathfrak{q}$ be two independent generic divisors of degree g, i.e. let them be such that $k(\mathfrak{p})$ and $k(\mathfrak{q})$ are free over k. There is a rational map $F : C \times \ldots \times C \to U$ of the product of C taken g times with itself generically onto U, such that $F(P_1, \ldots, P_g) = u$. As the g points $P_1, \ldots, P_g$ and the prime rational cycle $\mathfrak{p}$ over $k(u)$ determine each other uniquely, we shall also write $u = F(\mathfrak{p})$. Put $v = F(\mathfrak{q})$. As in the preceding proposition, there exists a unique positive divisor $\mathfrak{m}$ of degree g such that

$$\mathfrak{p} + \mathfrak{q} - \mathfrak{o} \backsim \mathfrak{m}$$

and if we take $\mathfrak{a} = \mathfrak{q} - \mathfrak{o}$ in that proposition, then we see that $\mathfrak{m}$ is a generic divisor on C of degree g, rational over $k(\mathfrak{p}, \mathfrak{q})$. If we put $w = f(\mathfrak{m})$, we must have by symmetry

$$k(u, v) = k(v, w) = k(u, w).$$

The normal law on U is then defined by putting $f(u, v) = w$, and the associativity is obvious from the consideration of the divisor class $\mathfrak{p}_1 + \mathfrak{p}_2 + \mathfrak{p}_3 - 2\mathfrak{o}$, taking $\mathfrak{p}_1, \mathfrak{p}_2, \mathfrak{p}_3$ generic independent divisors over k.

According to general results on group varieties, we can choose the model U to be a commutative group reflecting the law of composition we have just defined. We denote this group by J, and call it the Jacobian variety of C.

The above construction, entirely formal once we have the Riemann-Roch theorem, can be adapted to construct Rosenlicht's generalized Jacobians, whose definitions we briefly recall. Assume for a moment that C is complete but may have singular points. Let S be a finite number of points of C, containing the singular points, and assume that S is k-closed. Let K be the field of functions on C, and let K_S be the subring of functions in K obtained by taking the intersection of the local rings belonging to points of S. Let $D_S(C)$ be the subgroup of divisors of C generated by the points of C not in S. One can then define linear equivalence using only the units in K_S and one can prove a generalized Riemann-Roch theorem. If $\mathfrak{a}$ is in D_S, then we put $L_S(\mathfrak{a}) = L(\mathfrak{a}) \cap K_S$, $l_S(\mathfrak{a}) = \dim L_S(\mathfrak{a})$, and there is a formula

$$l_S(\mathfrak{a}) = \deg(\mathfrak{a}) + 1 - g_S + \delta_S(\mathfrak{a})$$

with a suitable constant g_S. The generalized Jacobian is then obtained by a construction entirely analogous to the one just described.

Returning to a complete non-singular curve, one can prove in addition that the Jacobian is complete, and is therefore an abelian variety.   The source of this result lies in the following proposition.

PROPOSITION 10. *Let C be a complete non-singular curve, defined over k, and let $\mathfrak{a}$ be a divisor on C. If $\mathfrak{a}$ is linearly equivalent to 0, then any specialization of $\mathfrak{a}$ is also linearly equivalent to 0.*

This proposition depends on elementary intersection theory, and we shall not prove it here. Granting its truth, it is a simple matter to prove that J is complete. We then get a rational map

$$F : C \times \ldots \times C \to J$$

of the product of C taken with itself g times onto J. By Theorem 4 of Chapter IX, this map decomposes into a sum of rational maps of C into J which by symmetry must be all equal, i.e. we get

$$F(P_1, \ldots, P_g) = \sum \varphi(P_i) + c$$

where $\varphi : C \to J$ is a rational map, and c is a constant on J. This map induces a homomorphism of the group of divisors of degree 0 on C onto the points of J, and one verifies without difficulty that the kernel of this map consists of the divisors which are linearly equivalent to 0. This is the abstract analogue of the classical theorem of Abel and Jacobi. For the proof, we refer the reader to Weil [5].

If $g = 1$, then we can apply the Riemann-Roch theorem directly to the curve itself, which is known as an elliptic curve.   The addition of the points on the curve coincides in the classical case with the addition formula derived from the addition theorem of elliptic functions, and topologically our abelian variety is a complex torus.

7. Harnack's theorem

We conclude the book with a proof of Harnack's theorem concerning real curves.   The only part of the Riemann-Roch theorem which we need is that if $\mathfrak{a}$ is a divisor on a curve, and $\deg(\mathfrak{a}) \geqq g$, then $l(\mathfrak{a}) \geqq 1$.

We consider a complete non-singular curve, defined over the real numbers R. The set of all points on C can then be topologized in a natural fashion, and constitutes a compact space, locally homeomorphic to the reals. It has therefore a finite number of components, each of which is homeomorphic to a circle. Harnack's theorem states that the number of such components is $\leq g + 1$.

Let B be one of the components. Consider the function field $R(C)$ consisting of functions on C defined over R. Let f be a function in $R(C)$. Then if P is a real point of C, $f(P)$ is a real number, or f has a pole at P. We contend that f cannot have a single zero of order 1 and no pole on B. Indeed, if f has just one zero of order 1 at the real point P on B, then $f(Q) > 0$ and $f(Q') < 0$ for Q and Q' close to P: Naively speaking, f changes sign as it runs through P. But $B - \{P\}$ is connected. Since f is continuous on $B - \{P\}$, we get a contradiction.

Now suppose there were more than $g + 1$ components. We would get at least $g + 2$. Select one point on each such component, and let $P_1, \ldots, P_{g+2}$ be these points. Let Q be a complex point, and Q^* its complex conjugate. Let $\mathfrak{q} = Q + Q^*$. The divisor $\mathfrak{a} = (g + 1)\mathfrak{q} - (P_1 + \ldots + P_{g+2})$ has degree g, and is rational over R. Hence there is a function f in $R(C)$ which is in $L(\mathfrak{a})$. Thus f has at most g more zeros in addition to $P_1, \ldots, P_{g+2}$, and has no pole on any of the real components. This implies that on at least two components f will have only a single zero, a contradiction.

Literature

Weil's proof of the Riemann-Roch theorem [3] is generalized by Rosenlicht [2] to construct generalized Jacobians [3]. Practically everything we know on abelian varieties (in the abstract case) can be traced to the theory of curves, via the Riemann-Roch theorem and the Jacobian.

Our proof that the sum of the residues of a differential form is 0 follows that given by Hasse [1].

Hasse, H.

[1] Theorie der Differentiale in algebraischen Funktionenkörpern mit vollkommenem Konstantenkörper, J. Reine u. angew. Math., Vol. 172, 1934, pp. 55—64.

Rosenlicht, M.

[2] Equivalence relations on algebraic curves, Annals of Math., Vol. 56, No. 1, July 1952, pp. 169—191.

[3] Generalized Jacobian varieties, Annals of Math., Vol. 59, No. 3, May 1954, pp. 505—530.

Weil, A.

[4] Zur algebraische Theorie der algebraischen Funktionen, J. Reine u. angew. Math., Bd. 179, Heft 2, pp. 1938, 129—133.

[5] Variétés abéliennes et courbes algébriques, Hermann et Cie, Paris, 1948.

INDEX

This index includes all the terms formally defined in the book, except that we refer once for all to Chapter IV, § 6 for the extension of local notions to abstract varieties.